Teaching Maths for Mastery

Practical strategies for primary schools

John Bee

BLOOMSBURY EDUCATION
LONDON OXFORD NEW YORK NEW DELHI SYDNEY

BLOOMSBURY EDUCATION
Bloomsbury Publishing Plc
50 Bedford Square, London WC1B 3DP, UK
Bloomsbury Publishing Ireland Limited
29 Earlsfort Terrace, Dublin 2, D02 AY28, Ireland

BLOOMSBURY, BLOOMSBURY EDUCATION and the Diana logo are trademarks of Bloomsbury Publishing Plc

First published in Great Britain 2025 by Bloomsbury Publishing Plc

This edition published in Great Britain 2025 by Bloomsbury Publishing Plc
Text copyright © John Bee, 2025

A catalogue record for this book is available from the British Library

ISBN: PB: 978-1-80199-610-5; ePub: 978-1-80199-613-6

2 4 6 8 10 9 7 5 3 1 (paperback)

Typeset by Newgen KnowledgeWorks Pvt. Ltd., Chennai, India
Printed and bound in Great Britain by TJ Books, Padstow, Cornwall.

To find out more about our authors and books visit www.bloomsbury.com
and sign up for our newsletters

For product safety related questions contact productsafety@bloomsbury.com

Acknowledgements

Writing a book on mathematics has been a journey of insights, challenges and invaluable support. I am deeply grateful to all those who have contributed to this endeavour.

First and foremost, I extend my heartfelt gratitude to Carl Newton, whose unwavering encouragement and support helped me to navigate the highs and lows of writing. Your love, understanding and kindness continue to inspire me.

I thank my mam Ann, dad David and brother David for your kindness and support and for believing in me all the time. Your love has no boundaries.

Thank you to my best friends, Lynsey MacGregor, Natalie Moffat and Jennifer Petch. You ground me and make me laugh.

I would like to express my sincere gratitude to special colleagues who have helped me along the way. Your expertise, experiences and passion have helped to shape who I am.

Claire Brown, for allowing me to live and breathe the pages of this book each day – I can only dream of being as knowledgeable as you are. Emily Pringle, for being a great colleague and offering valuable support and challenge on mathematics (and a wonderful chapter of this book). Nick Hurn and Lucie Stephenson, for giving me wonderful opportunities. Colleagues including Barbara, Andrea, Martine, Graham, Jon, Paula, Imelda, Amy, Anthony and the whole of the education team and wider BWCET team. Laura Tullock (who proofread chapters of this book and whose passion and love of maths continues to inspire me), Lisa Heatherington and Laura Greener. The whole of the *GREAT* north team!

Finally, I would like to thank the hundreds of children I have taught in my career. You are at the heart of this work and your curiosity and enthusiasm are written across the pages of this book. Your confused but inquisitive faces and occasional 'aha' moments remind me that mathematics, like life, is all about the journey, and not just the destination. You have inspired me to strive for excellence in teaching.

To everyone who has contributed to this book, whether intentionally or not, thank you for adding your unique colour. Here's to embracing the joy of numbers and unlocking the mysteries of life and the universe – one (small, coherent) step at a time!

Foreword

Claire Brown, Senior Director for Performance and Standards at Bishop Wilkinson Catholic Education Trust

The first time I met John Bee, I was struck by his absolute passion for mathematics, but also by his resolute determination that every child should benefit from an exceptional mathematics education to enable them to flourish. I was delighted that we appointed John to lead maths improvement across our 42 primary and middle schools at Bishop Wilkinson Catholic Education Trust (BWCET). A year on and the difference that he has made to our education team and our schools is significant.

This impact has been driven by the concepts that are central to this excellent book. At BWCET, we are intensively focused on the quality of our curriculum. Thinking deeply about what we want our children to learn, know and retain over time is critical. And this must be the first step before we decide how to teach that crucial knowledge. We cannot start the teaching process by planning an attention-grabbing activity that will capture children's interests before we think about what it is that we want children to learn. Put simply, that places the cart before the horse.

This book helps educators to think deeply about the mathematical knowledge that is most crucial for pupils, either because it's worth everyone just knowing it or because it establishes the foundations on which pupils will build when they encounter future learning. John emphasises how essential precision-planning is in mathematics, considering those small steps that make up key elements of more complex knowledge. He challenges us to think carefully about our non-negotiables: the things that are essential for children to know, whether that's accurate number formation by the end of the Reception year, because we know it becomes a barrier for pupils when they're in Key Stage 1 if this is not the case, or automaticity in number facts to free up pupils' working memory to solve complex problems.

The guidance that John offers here to teachers is perfectly focused on how we ensure that teaching supports pupils to truly master the essential knowledge set out in our curriculum. Gimmickry and popular short-cuts may provide short-term success, but they set pupils up for a lifetime of challenges with maths. I speak from personal experience as someone who, on paper, in terms of exam performance, appears to have been extremely successful in maths! It was only many, many years later, as a headteacher in Wallsend, as I and my excellent team grappled with breaking through a plateau in pupils' maths achievement, that I recognised how damaging those mathematical 'tricks' and short-cuts can be. This superficial learning limits children. In this book, John instead illustrates how lessons and teaching sequences can be designed to ensure that our pupils deeply understand and retain mathematical knowledge, rather than simply encountering it, recognising that pupils' enjoyment of mathematics arises from the confidence that they feel when they really 'get it' and experience success, rather than just because a lesson is 'fun'.

John's personal commitment to ensuring the highest expectations for all children, regardless of their background or starting points, shines through this book. He challenges us to consider how some of our common practices may put a glass ceiling on the achievement of our pupils with special educational needs or those from a disadvantaged background. He is uncompromising in his ambition, whether it is in the mathematical vocabulary that all pupils should learn, the access that every pupil must have to more complex problem-solving to apply their mathematical knowledge or the opportunities to articulate their mathematical thinking.

Whether you are a headteacher or a subject lead seeking guidance to help you on your school improvement journey in mathematics, or a class teacher focused on improving your own practice, this book will provide you with both the inspiration and the practical strategies that will enable you to provide a high-quality mathematical education for the children you serve.

Contents

Introduction

I have been teaching for 14 years at the time of writing this book. Since I started, I have seen the transition from the National Numeracy Strategy through Big Maths, Problem Solving Friday and everything in between. Looking back, I wince at some of my maths teaching. I've since realised that maths has nothing to do with crocodiles when comparing numbers, bus stops when dividing or unrelated rhymes when rounding numbers. Instead, I now favour deep conceptual understanding of how and why mathematical concepts work. In other words, I want children to master concepts.

I visit many schools as part of my job as a school improvement adviser, and I see varying degrees of success in implementing teaching for mastery. This book aims to support your thinking around effective and practical ways to do this in your own classroom, across your own school and across schools.

Some of my thinking, resources and materials are available from www.mrbeeteach.com. I started sharing resources online in 2019, and since then I have developed a maths curriculum and resources to support teaching for mastery. I love to see and hear how they are being used in schools. Please do share them on social media @mrbeeteach.

Let's demystify teaching for mastery. I often think of it as teaching until mastery – ensuring that all children have a secure understanding of what is being taught before progressing. It puts ambition for all at the centre of our beliefs and assumes that all children can achieve in maths, regardless of perceived ability. It is equitable and rejects the notion of labelling children as 'less able' or 'more able'. Instead, it operates on the assumption that all children can achieve. We want all children to have a great depth of understanding. More time is spent on key topics, rather than rushing through a wide range of topics, and the focus is building on strong foundations. The ultimate goal is for all children (with limited exceptions) to acquire facts and knowledge from their year group and key stage.

The five big ideas of teaching for mastery

According to the National Centre for Excellence in the Teaching of Mathematics, there are five big ideas of teaching for mastery (NCETM, 2017).

- coherence
- representation and structure
- mathematical thinking
- fluency
- variation.

Teaching for mastery

Teaching for mastery is a teaching approach that promotes deep understanding of mathematical concepts. It builds lesson on lesson, unit to unit and year on year to allow children to make connections between concepts and fully understand how and why concepts work. This is done by providing students with opportunities to practise and apply concepts in a variety of ways until they have mastered them.

Many schools in England either are teaching for mastery or have heard of teaching for mastery, but what does it actually mean? Mastery means having high expectations of *all* children. This may be obvious to some, but there was a time in classrooms across the United Kingdom where teachers would deliberately plan activities based on preconceived ideas of children's abilities. What this meant in reality was that those children who were considered 'least able' were given the least demanding, often procedural activities to complete, and those children considered 'most able' were given the most cognitively complex activities. Lessons would be littered with many different tasks, pitched at what the teacher thought was appropriate (not taking into account the fact that some children may be more confident at multiplication but less confident with measuring or other areas of the maths curriculum). Grouping by perceived ability creates huge gaps and can be extremely damaging. This approach in school would happen deliberately and systematically year after year, would be insisted on by senior leadership teams and was often considered effective practice. What actually happened was that the gap in progress and attainment widened as children progressed through school. It also led to cultures of low expectations for particular groups of children.

I still hear of some classrooms offering 'mastery challenges'. This completely misses the point of teaching for mastery. Mastery is an approach, a mindset and a way of thinking about teaching. Mastery is not a label for some children to progress on to when they have become confident with fluent memorisation of some of the methods from your curriculum. Mastery is for all. We want all children to achieve in maths. We want all children to master the curriculum.

This model assumes that maths is only for some children. Indeed, some teachers may have thought that their job was to check which children could and could not do maths and then teach those who could. Our thinking, thankfully, has moved on, and mastery assumes that everyone can learn maths at a deep level.

Retrieval

Retrieval is at the heart of mastery. In maths, concepts tend to build logically and coherently. The skill of a teacher here is not to see who can and cannot do maths; instead, the focus is on finding starting points. In some cases, it is wholly appropriate to track back several years as a starting point and build the concept to remind children of what they already know or have learned. We might consider this as 'stage not age'. For example, when teaching children to add fractions with different denominators in Year 5, we might track back to Year 3 or 4 and discuss what fractions are (parts of wholes) and perhaps practise counting forwards and backwards in fractional steps. We might also question what a numerator or denominator is, before practising with adding unit and then non-unit fractions with the same denominators, and then a refresh of equivalent fractions. This may take several minutes, depending how well-remembered it is among children in your class, or it may take several lessons to reteach and remind children. Children forget stuff. Adults forget stuff. It's how we learn. After all, memory is the residue of thought(s). The point is that all children are on the journey with you and you meet them at their current destination. Figure 0.1 is an example of 30 children in a Year 5 class; each child is represented by a dot, to show the differing starting points. The line suggests where we may begin to ensure that all children have opportunities to revisit and remember what they have been taught. Anyone behind the line may need some pre-teaching or intervention to engineer success.

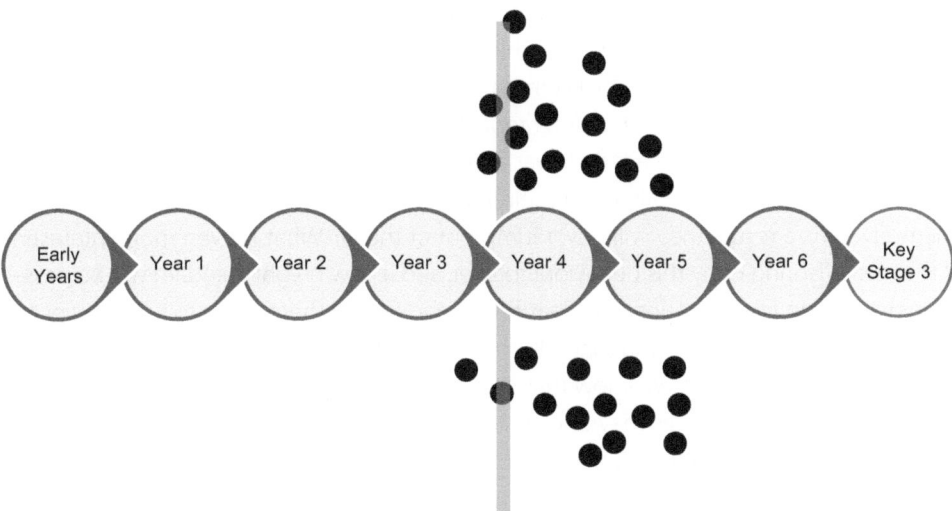

FIGURE 0.1 Example representation of the different attainments of children in a class

With a class of 30 children, you have 30 different starting points and 30 different experiences of mathematics from home, online and other external influences. If we assume that all children have mastered content to their given age, we can soon run into difficulty – ask any Year 6 teacher, who, out of necessity, tracks back to previously taught concepts so that they can access learning in Year 6. This is fairly standard practice, but this practice is perhaps because children have not mastered concepts deeply enough in their year groups before progressing to the next academic year. I am often struck by an 'oh well' culture in schools. What I mean by this is that the children in Year 1 may not have mastered some elements of number or place value, but teachers have pressure to teach the full curriculum, so they think 'oh well' and move on, rather than spending more time on priority areas of the maths curriculum. When children arrive in Year 2, the same thing happens, as it takes the Year 2 teacher even longer to reteach concepts with which children were not secure in Year 2, and then 'oh well', they made some progress, but bits of the wider maths curriculum may be a bit hazy. By Year 3 – and you can see where I'm going with this – children become confused, and so mindsets around maths start to solidify and become fixed. It's easy to see how this soon translates to 'I can't do maths' or other similar phrases that we may have heard. We need relentless focus on ensuring that children understand key concepts across the curriculum. I look at the curriculum more as a set of promises that we make to the next teacher as to what children will know.

Environment: The flea in a jar experiment

When I first started teaching, I remember sitting in a professional development session with a local authority behaviour adviser (remember those?). She began the session by telling us about an experiment. I'm still not entirely sure whether this experiment is real or not, but the idea behind it is powerful. This experiment placed a number of fleas in a jar. Fleas can usually jump to around 100 times their body size. The experiment places the fleas inside of a jar and places a lid on it. After repeatedly hitting the closed lid, the fleas learn to jump a little lower and then a little lower again, until they no longer hit the inside of the lid. Days later, the fleas have learned their limits and now jump so that they do not hurt themselves. As a result, they will never jump out of the jar. What is even more interesting is that flea offspring copy this behaviour forever, and a new height standard will be passed from generation to generation. The fleas have conditioned themselves to jump lower and have assimilated a new schema that tells their brains that it will never be possible to jump further than a given point. We know that humans have the advantage of being able to act and think rationally, but this may also be a disadvantage when what we see around us, who we communicate with and how we communicate with each other, and what we are told and taught become our lived reality. In short, our environment shapes who we are and how we think.

If schools are deliberately and systematically grouping children by perceived ability or by any other measure, we may be inadvertently placing a lid on how far they can jump.

Teaching for mastery attempts to unpick that and offers a vision of equity to maths education. High expectations are placed on *all* learners, and differentiation is achieved through the depth of their understanding and ability to make connections with the mathematics that they are using. It works towards the idea that all children could and should understand maths. For example, all children in a class may be learning that fractions are a part of a whole. When teaching for mastery, we would want all children to understand this concept. Once secured, it may be deepened and strengthened by offering structured activities that deepen children's understanding. (For example, you might use a map of the UK to show that England is part of the whole UK, and then apply the same idea to a map of Europe, where the UK becomes part of that larger whole.) Knowing this well allows children to consider the concept of what fractions are (parts of the whole) before they have really looked at how to write a fraction. Notice that there is no comment on giving the children considered 'least able' something different or the children considered 'most able' something from another year group, or rushing through to another concept. In fact, when I wrote those words, they seemed very archaic and out of place, yet I still hear them regularly. Labels in this way are not only unhelpful but also counterproductive.

Six big ideas

I mentioned five big ideas of teaching for mastery earlier. There are probably six in reality. I realise that this is a maths book, and it now appears that I cannot count! The extra one runs as a thread through all of the big ideas and focuses on vocabulary. Maths has its own specific vocabulary and it cannot be left to chance for children to use words such as 'denominator', 'quotient' or 'subtrahend' (more on these later). These principles should be considered in every lesson, rather than moving through them at different stages within lessons or across lessons. Effective lessons include elements of coherent structure (including small steps), representations to reveal mathematical structures, mathematical thinking, varied fluency, variation theory and specific and technical mathematical vocabulary.

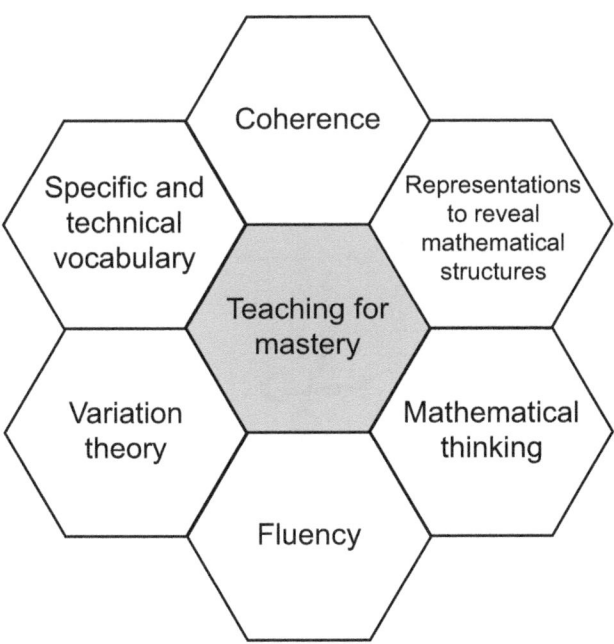

FIGURE 0.2 The five big ideas of teaching for mastery (and an additional one specifically on vocabulary)

This book uses these ideas as a framework and considers how the concepts may be taught. It will consider each idea discretely and then examine how they connect to the other ideas, to create a teaching for mastery approach.

Coherence (and consistency)

The mathematics curriculum should be taught in a coherent way, with each concept building on the previous ones. This helps pupils to see the connections between different mathematical ideas and to develop a deep understanding of the subject. The National Curriculum is designed in distinct year groups, and concepts build year on year. As concepts build in a logical and sequential way, we can use them to track back and revisit previously taught concepts to ensure depth of understanding before new learning is introduced.

Representations to reveal mathematical structures

Pupils should be given opportunities to use a variety of representations to help them to understand mathematical concepts. This could include using concrete objects, pictures, diagrams, symbols and words. Using different representations helps pupils to make connections between different ways of thinking about mathematics and to develop a deeper understanding of the concepts. For example, we may tell children that to calculate the area of a triangle, we multiply the sides and halve the answer, but this does not really explain why or how this may be taught. Instead, if we have a card rectangle cut out for each child that measures 5 cm x 6 cm and show children that the area it covers is 30 cm², demonstrating how this may be cut in half to leave a triangle reveals why this works, as shown in Figure 0.3.

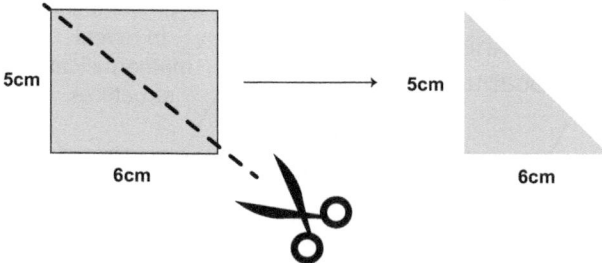

FIGURE 0.3 The structure of the area of a triangle being half of the area of a rectangle

Thinking mathematically

Pupils should be encouraged to think mathematically by asking them questions that require them to use their reasoning skills. These could include asking them to justify their answers, to explain their thinking or to solve problems in different ways. Thinking mathematically helps pupils to develop a deeper understanding of the concepts and to become more independent learners.

Varied fluency

Pupils should have opportunities to practise mathematical skills in a variety of ways. This helps them to develop fluency in the skills and to become more confident in using them. Varied fluency also helps pupils to see the connections between different skills and to develop a deeper understanding of the concepts.

Variation theory

Variation theory is a way of thinking about how pupils learn new concepts. It suggests that pupils learn best when they are presented with a variety of problems that require them to use the new concept in different ways. Variation theory helps teachers to plan lessons that will help pupils to develop a deep understanding of the concepts. We may use **conceptual variation** or **procedural variation** to draw attention to the essential and non-essential properties of a concept.

Conceptual variation

The example in Figure 0.4 demonstrates the concept of fractions but draws attention to the conceptual understanding of fractions as being *equal* parts of a whole.

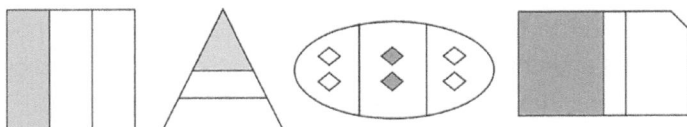

FIGURE 0.4 Conceptual variations of a third, featuring some examples and some 'non-examples' (misconceptions) of what a third is

Procedural variation

An example of procedural variation is the connection between **4** x 6 and **8** x 6, which reveals the structure of how doubling one factor doubles the product (the answer). Similarly, there are other links, such as 3 x 7 and 7 x 9, that reveal that trebling one factor trebles the product (this also shows the commutative nature of multiplication). These examples show how we keep one thing the same and change another. Look at how carefully the numbers have been chosen to guide children and allow them to make connections and links between the mathematics that they are using.

Specific and technical maths vocabulary

Pupils should be taught the specific and technical vocabulary that is used in mathematics. This helps them to communicate their mathematical ideas more effectively and to develop a deeper understanding of the concepts.

Using terms such as 'dividend', 'divisor' and 'quotient' with division allows for greater depth to reasoning and avoids clumsy descriptions of what is happening. It can unlock a world of understanding and allow children (and teachers) to make sense of maths. For example, look at Figure 0.5.

$$
\begin{array}{cc}
3\,1 & 3\,1\,r3 \\
4\,\overline{)\,1^12\,4} & 4\,\overline{)\,1^12\,7}
\end{array}
$$

$$
\text{Quotient} \qquad 3\,1\,r1 \qquad\quad 3\,2
$$
$$
\text{Divisor}\,\overline{)\ \text{Dividend}} \qquad 4\,\overline{)\,1^12\,5} \qquad 4\,\overline{)\,1^12\,8}
$$

$$
\begin{array}{cc}
3\,1\,r2 & 3\,2\,r1 \\
4\,\overline{)\,1^12\,6} & 4\,\overline{)\,1^12\,9}
\end{array}
$$

FIGURE 0.5 Short division demonstrating that a remainder can never be more than or equivalent to the divisor

Here, it is not difficult to see how children could arrive at generalisations that the remainder may not be greater than the divisor. It may allow for conjectures, such as: 'Would 445 have a remainder if the divisor was 4? How do you know?' This generalisation, through the use of specific and technical mathematical vocabulary, can be a powerful game-changer for teachers and children. You can see how quickly this may allow children to become fluent and clear. The coherent design of this activity reveals the structure through procedural variation. This interplay between the big ideas shows teaching for mastery in all its masterful glory.

Teaching for mastery can be a challenging but rewarding approach to teaching mathematics. It requires teachers to have a deep understanding of the mathematics and to be able to plan lessons that will help pupils to develop a deep understanding of the concepts. However, the benefits of mastery learning are clear. Pupils who are taught using a mastery-focused approach are more likely to develop a deep understanding of mathematics, to be more confident in their mathematical skills and to be more successful in their future studies. The following chapters will exemplify how you too can master the art of teaching mathematics.

1 Key concepts

Curriculum prioritisation

The National Curriculum is overloaded with content. Knowing which areas of maths to prioritise is important in order to ensure that children achieve well in maths. This is particularly important in the time-strapped school day and with conflicting priorities. Prioritising fundamental concepts that link to the 'wider' maths is an important consideration. These building blocks are essential to build understanding of other concepts later on. For example, learning to count in multiples of 2, 3, 4, 5, 10, 25, 50, 100 and 1,000 may help when reading charts and graphs during lessons on statistics. But which concepts should be prioritised and why? Place value, number facts, the four operations and fractions (plus a little geometry knowledge) are priority areas, as they are the foundation of mathematical understanding and connect to much of the rest of the maths curriculum. When teaching decimals, we need knowledge of place value, using them in calculations or converting them to fractions, so those overarching areas of maths need prioritising.

Connections are everywhere across the maths curriculum. However, connections cannot be made without strong foundational knowledge. For example, In Year 1, counting in multiples of 2, 5 and 10 will help children when unitising coins later or applying this understanding to scales charts and graphs when thinking statistically. Without this foundational knowledge of counting in 2s, 5s and 10s, acccess to coins and statistics may be somewhat more challenging.

I think some schools do this rather well already. I often see 'SATs clubs' or interventions that tend to be focused around understanding of number (place value, four operations or fractions). Even if their rationale isn't clear, it seems obvious to me that a deep understanding of these concepts is useful for applying them to other concepts in the wider maths curriculum.

Leading sustainable change

Much of this book relies heavily on leadership buy-in – developing a shared vision, culture and set of principles to support teaching for mastery. We need to consider how policies, practices and systems support the approach and how embedded, collaborative professional development should foster professional learning as part of school development. We must also consider what this means for all stakeholders within a school and our roles within a school in doing that.

Sustained and focused professional development is better than a randomly placed staff meeting. Staff meetings can tend to be didactic in approach, and in my experience can be an elaborate 'to do' list rather than professional development. They are often at the end of a busy school day, when attention has turned to a pile of marking to be done or what to cook when people get home. Staff meetings are just one way in which we can lead change in schools. Developing people, their thinking, skills, beliefs and pedagogy takes time. Professional development can be developed across school(s) to focus on core activities such as collaborative planning, observing learning (with specific foci, support and challenge for leaders) and analysis of lessons and teaching approaches.

These professional development approaches are by no means exclusive to maths; in fact, developing them across curriculum areas or departments can be transformative and lead to sustained change and a culture of excellence.

Communities of excellence

This has been the part of professional development that has had the biggest impact on my practice. Many years ago, when maths hubs were being established across England, I invited groups of up to 15 teachers to observe my lessons. I would discuss the sequence of the lesson, prior learning and what I wanted the children to learn in the current day's session. I'd then teach a lesson and afterwards we would have post-lesson analysis, framed around some principles: coherence, representation and structure, mathematical thinking, fluency and variation. All comments would be wrapped around these ideas and we'd discuss the impact of choices made in lessons. This significantly helped the process of change across school, as it acted as a model of what could be possible and identified areas of priority throughout the school. It also acted as subject knowledge development and allowed me, and my school, to work collaboratively with other schools. What's more, once I and other colleagues became more and more confident with our approach, we'd invite other stakeholders into lessons, such as governors, parents, teaching assistants, parallel classes and, once, a local MP, to learn, grow and collaborate.

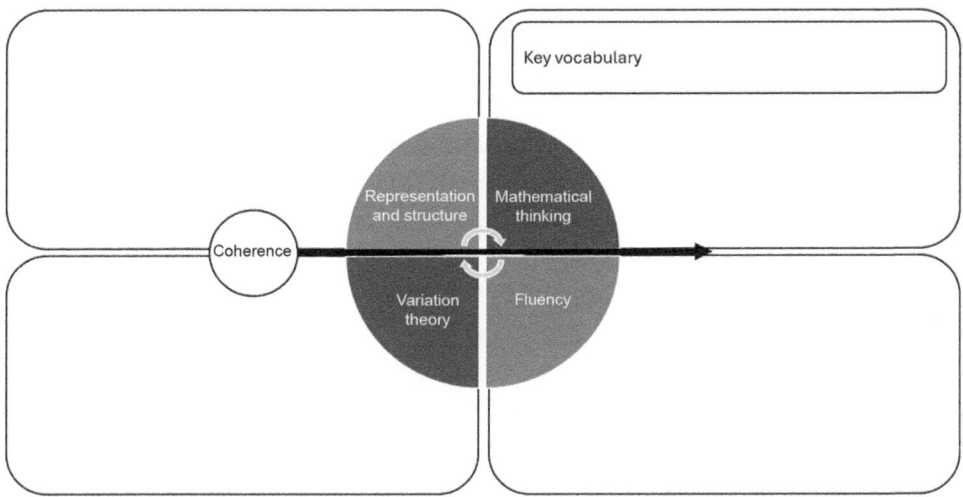

FIGURE 1.1 Suggested lesson observation form using a modified NCETM 'five big ideas' to include a focus on specific vocabulary (adapted from NCETM, 2017)

Figure 1.1 shows the lesson observation notes form that I would give to those who were watching the teaching. It is just something simple but it allowed a clear framework of how to observe the teaching.

Collaborative planning

I started planning collaboratively with other teachers many years ago. We'd pore over the detail of a session and what we wanted children to attend to. This really sharpened my thinking around what I was teaching and, importantly, how I would teach it. We'd consider the sequence and structure of lessons, how to demonstrate a representation to reveal the structure being taught and what children might do as activities. We'd discuss the vision for maths, vocabulary, how to get children to reason and solve problems, and how we might present material and in which order. The result was transformational in the schools in which I worked. A clear vision and shared understanding of the approaches, vocabulary and culture of maths was developed. It was magical. In China, this is called 'lesson design' rather than lesson planning. Designing the architecture of learning through collaboration with others can support all involved with maths teaching.

Staff meetings

Staff meetings then became less of an elaborate to-do list and more of an opportunity to develop thinking and collaborate on areas that we needed to prioritise. That may have been coherence and small steps or perhaps a focus on teaching specific vocabulary or teaching with variation. I would often record lessons or parts of lessons and use these to analyse teaching (in a non-judgemental way). We would shine a spotlight on part of the lesson and form a discussion around it. It could be the use of mathematical vocabulary, how representations were used or something else, depending on the intended purpose of the staff meeting.

Approaching professional development in this way started to change the fabric and culture of the schools in which I worked. Colleagues were more open to learning and development and would ask questions of each other. Subject knowledge built as a result, colleagues were motivated to improve and try out different strategies, and they reflected on pedagogy and embedded the vision into their practice.

Barriers

Of course, there were struggles throughout this time. Things that could be tricky included releasing staff, initiative fatigue setting in, finding time for meetings, getting senior leader buy-in, staffing changes and more. Releasing teachers and finding time tended to be the main barriers that I encountered. From my experience, constantly asking teaching assistants to cover regularly for whole maths lessons while maths teachers undertook CPD (continuing professional development) led them to disillusionment and even resentment. Instead of this, I would invite teachers to watch part of a maths lesson with a very specific focus (perhaps on representations) and then feed back at the end of the day for ten minutes. This was bite-sized professional development, but it also meant that it could happen regularly and that change would be sustained. Thinking creatively can lead us to wonderful solutions.

Visualise it

In my first years of teaching, I can remember using presentations and talking through them carefully. I'd set children off to do a task and I'd often finish in despair at what they had produced. I spent a while using large squared paper taped to a small whiteboard to model activities with children, until one day I used the camera app from a well-known tablet device. I had suspended it on a stand so that the board was displaying the table below the tablet. I started to use my own maths book too. I'd stick in some of the tasks that I wanted children to do during the lesson and I'd model how I might approach them. This may have been using written reasoning, key vocabulary, models, images or a mixture of all of these. This worked in a few different ways. It gave children an explicit worked example but it also took care of the operational side of things, such as how to set ideas out or how to draw number lines or bar models that accurately related to the task or problem on which they were working. Essentially, I was unpicking the structure of problems with children. Once I'd worked through one, children would be given problem(s) that were structurally similar to the one that I had shown them so that they could practise. I didn't realise it then, but I was essentially adopting an 'I do, we do, you do' approach to teaching maths.

For children who finished quickly, or if I wanted to strengthen and deepen understanding, I could write a few more questions in my maths book for them to work on, perhaps changing numbers or the structure of the problems, so slightly using variation theory to develop their thinking further. It all depended on what I was teaching, and teaching in this way requires a strong and robust understanding of mathematical concepts and how they link together and connect.

Later down the line of using a tablet, I took another leap into an idea that was a game-changer. I'd been using my own maths book more and more to model my thinking and working out. I want you to imagine the power of me facing the wall, with the interactive board behind me displaying the squared page of my own maths book. Despite teaching in this way for parts of the lesson, some children still needed additional support. And here's what changed the game again. I began to press record as I was modelling a concept or a process. I was able to play the recording back on loop, so that children could follow a process again and again while I was working with groups at different places in the classroom.

If I was working with a group, I'd take my tablet device and work with a group using my own maths books to support with some of the questions, while those not in my group could see the screen on the board and be reminded of the expectation, concept or process. It was a really versatile tool and one that the children really liked, as it allowed all children to experience success. One particularly challenging class I had was so engaged with this that some children told me how I could replay the videos but on double the speed or how to slow it down to half the speed. All of these approaches felt like having an additional teaching assistant in the classroom at all times. I also started this approach across the whole curriculum, with huge success.

I would often revisit concepts from across Key Stage 2 to check children's understanding. If children had forgotten or needed more practice, I had a reliable bank of ready-to-go explanations that we would use as a guided practice at the start of lessons for retrieval. That then freed me up to speak with children or groups of children about work completed yesterday, do a bit of pre-teaching before the day's session or sit back and just observe to assess children's confidence.

I'd sometimes pre-record procedural explanations before lessons using this approach. I'd then introduce the idea and play the video. This meant that I could then sit at a table and support children's thinking step by step, perhaps using manipulatives or checking that they were setting calculations out accurately. I'd also loan the tablet device to teaching assistants, who would watch the videos to remind themselves of concepts. This meant that there was parity between the methods that I was teaching and the methods that teaching assistants used. I am sure that there are many other ways in which this approach could be used, adapted and enhanced. Using this tool allows for teaching until mastery has been achieved, by engineering high levels of success in classrooms.

Cultures of excellence

Developing a vision is the starting point. We might start by asking big questions, such as why it is important that maths is taught well in school and what problem we are trying to address. For me, I wanted to make maths accessible to all children. I wanted to undo the trend of lower-attaining children always being given activities and interventions that only focus on fluency. Now, fluency is very important, but it should sit alongside an understanding of the structure of mathematics, rather than learning things purely by memorisation (although even this type of fluency does have a part to play with times tables, number bonds, etc…). I wanted them to experience a full and comprehensive maths curriculum where they could reason and solve problems. I wanted all children to have a deep understanding of mathematical structure. I used to want equality, but this sometimes confused matters and meant that all children were given the same. What I actually meant was that I wanted equity: fairness for all children to achieve well in maths. Equality gives all children the same ambition, whereas equitable teaching meets children where they are currently and scaffolds them to develop their understanding. In other words, I want all children to master maths.

FIGURE 1.2 Equality versus equity

We know that the quality of teaching directly correlates to progress and attainment. We need to connect subject knowledge with pedagogical knowledge so that teachers can teach most effectively. Teaching for mastery is an approach designed to do this and is an implementation model to promote equity in maths. But what does it mean to master something? If we have mastered something, we can do it well, it's automatic and we can often show somebody else how to do it. We want this in our schools for maths too. Introducing and developing the five big ideas of teaching for mastery can drive long-term sustainable change.

Memory

Teaching until children have mastered concepts requires children to remember more. This may be through overlearning and retrieval approaches to ensure that learning is sustained and committed to long-term memory. Remembering the small minutiae of the whole curriculum may be a daunting task, but perhaps focusing on priority concepts is a good starting point. Using the curriculum priority 'ready-to-progress' criteria (DfE, 2020) can be a good frame of reference for which knowledge and concepts to prioritise. Revisiting concepts from previous year groups to consolidate and remind children at the start of lessons is time well spent. Key facts and knowledge should be identified, such as which times tables (and in which order) should be taught in which year group. When we are focused on teaching one small step, that means that we are not teaching all of the other steps from across our maths curriculum and we forget. We know that we need to forget in order to remember, and so deliberately planning what, when and how we want children to remember key knowledge is an important consideration. Figure 1.3 shows the 'forgetting curve', which demonstrates how quickly we forget things.

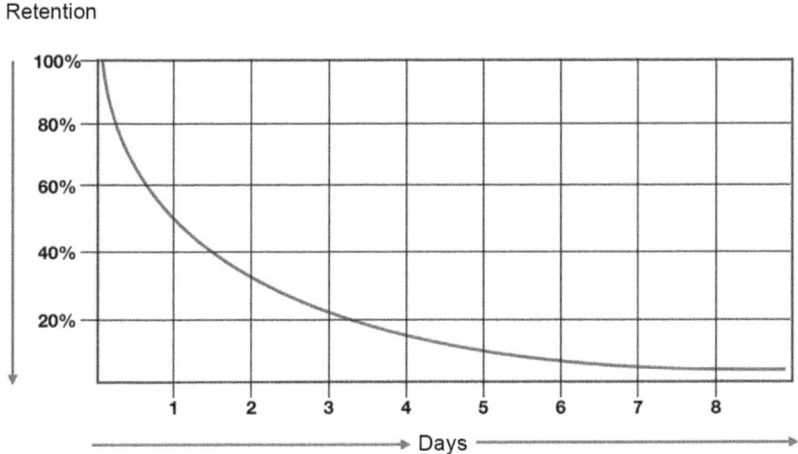

FIGURE 1.3 The forgetting curve (based on Ebbinghaus, 1885)

It is impossible to revisit all of the maths curriculum across a year when you have new content to teach, but prioritising key areas of your curriculum – place value, four operations, fractions and geometry – is a good starting point. Revisiting concepts from previous years and building on them, reminding children of recently taught concepts and prioritising what you want them to know are essential. In the long term, children will develop automaticity with key concepts, and so cognitive load will be reduced when they are solving problems. For example, prioritising times tables recall in Years 3 and 4 is important so that children can calculate area accurately and quickly, rather than stumbling over the recall of times tables facts in tasks. Similarly, prioritising automaticity of facts to 9 x 9 is even more important so that children can solve short multiplication questions (as they'll not be asked to multiply by two digits until upper Key Stage 2).

Assessment

All teaching informs assessment and all assessment informs teaching. When teaching until mastery, we want the vast majority of children to achieve high rates of success in lessons so that they are ready to progress to the next step. We might want to consider how we check that children can recall key facts, such as bonds to 10 in Key Stage 1, doubling in Reception or times tables in Key Stage 2. How do we measure this? By accuracy or speed? And does measuring and assessing in this way allow automaticity?

We might also consider the use of diagnostic or pivot questions in our classrooms, deliberately designed to draw out key thinking and misconceptions. For example:

$$\frac{1}{2} + \frac{1}{9} = \frac{?}{18}$$

- **A:** 2
- **B:** 3
- **C:** 10
- **D:** 11

Each of the possible responses reveals a potential misconception and allows teachers to pivot and adapt teaching in lessons to include more practice on a skill. For example, if children were to answer as follows:

A: 2 – we might deduce that they have simply added the numerators together.

B: 3 – they may have converted $\frac{1}{9}$ into $\frac{2}{18}$ and added the numerators to arrive at 3.

C: 10 – they may have converted $\frac{1}{2}$ into $\frac{9}{18}$ and added the numerators to arrive at 10.

D: 11 – they have got it correct.

However, we need to be mindful that some children may have put the two digits from the numerators together to arrive at the answer 11, and so we may need to provide another example. Another example could be:

$$\frac{4}{5} - \frac{1}{2} = \frac{?}{10}$$

- **A:** 3
- **B:** 5
- **C:** 12
- **D:** 4

Again, this question can be revealing as to which answer children choose.

A: 3 – this shows the correct answer by converting the calculation to tenths (but children may have simply subtracted the numerators and stumbled across the correct answer).

B: 5 – this may indicate that children have added together the numerators and ignored the operation.

C: 12 – this may suggest that children have added together all of the digits from the fractions (but not the 10 from the answer).

D: 4 – this could suggest that children have multiplied the numerators together (as this appears to work when we multiply the denominators together).

These types of assessments can take a long time to develop and design, because the answers need careful consideration as to which misconception they are going to draw out. Their careful design means that teachers can accurately diagnose where the misconception(s) are and identify how to resolve them.

Models of professional development

Professional development has moved on considerably since I started teaching. One-off courses were commonplace and we'd diligently return to school and try to emulate what we had been told or what we had seen. This has changed significantly in favour of a more long-term vision that can be sustained.

Beliefs are often the strongest in maths. Maths runs deep with people. Everyone has experience of being taught maths, whether that's been effective or not. Teachers can often rely on how they were taught concepts, which may revert back to gimmicks and tricks to get children 'over the line'. An equitable education in maths considers ability grouping (mixed-ability groupings), adaptive teaching and a coherent and logical sequence of lessons that enables all children to achieve. In the UK, we have been systematic and brutal at labelling children 'top', 'middle' or 'bottom' ability. If we do this, we bring our own biases to each and every lesson and limit what children can achieve. Of course, we will always have children who attain differently, but perhaps our ambition for all children should be that they have a great depth of understanding of maths concepts, how they work and how they connect to each other.

Even the label of 'greater depth' can become problematic and lead to abrasive attainment groups, where those children considered 'greater depth' get more opportunities to apply their understanding and learning. I think that it is the system itself that creates this systematic and unapologetic thinking among teachers and leaders, as that is ultimately how children and schools are measured. My argument is about high ambitions for all – for all children to receive a maths education that is rooted in sound conceptual understanding and for all children to have a connected understanding of how maths works.

Being labelled a 'low attainer' or, worse, 'lower ability' can be pretty grim in day-to-day life at school. Not only do children quickly realise that they're being given easier tasks in lessons, but they're then taken out of lessons that they enjoy to complete interventions. What's worse, the intervention tends to focus on fluency and the mechanics of unrelated, abstract methods delivered by teaching assistants. This, in turn, limits the opportunities that children have to reason and solve problems. We also know that this experience for children tends to impact those from the most disadvantaged of backgrounds. But there is another way – an equitable curriculum that prioritises key knowledge and constantly tracks back to prior learning of priority knowledge to ensure that it is committed to long-term memory. Essentially, we are teaching *until* mastery. Intervention can happen within lessons with the teacher, and will be flexible around what children seem to be struggling with on that day. Same-day intervention may be designed around those priority areas and revisit previously taught concepts, so that pupils can use and apply them in subsequent lessons. These small, coherent steps build up the tapestry of learning and allow for real depth of understanding.

Models of professional development are our best bet at developing teaching and the experience that all children get in schools with maths. We want high-quality teaching to be an everyday occurrence for children. This requires a tight vision from the maths leader and development of both pedagogical understanding ('how to') and content knowledge ('what to') of mathematical concepts. Teacher beliefs are central to this. Imagine that you are trying to implement teaching for mastery, but some teachers miss out on one of those two concepts because they do not understand what it means or think that they do not need to use it in their teaching or design of lessons.

TABLE 1.1 The impact of the absence of one of the missing big ideas

	Fluency	Mathematical thinking	Coherence	Variation	*Shallow understanding and reliance of teaching in the abstract*
Representation and structure		Mathematical thinking	Coherence	Variation	*No recall of key facts*
Representation and structure	Fluency		Coherence	Variation	*Limited vocabulary and lack of structure*
Representation and structure	Fluency	Mathematical thinking		Variation	*Disjointed and unconnected ideas*
Representation and structure	Fluency	Mathematical thinking	Coherence		*Inflexible use of knowledge*

To implement our vision for mastery and have ambition for all children, our professional development needs to focus on building knowledge, motivating teachers, developing techniques and embedding practice. We may make use of the 'implementation model' from the Education Endowment Foundation (EEF) to focus our thinking on what we want to develop across school (Sharples et al., 2024). The EEF suggest 'explore', 'prepare', 'deliver' and 'sustain' as a cycle of professional development.

TABLE 1.2 Simplified EEF implementation cycle (adapted from Sharples et al., 2019, p. 8)

IMPLEMENTATION PROCESS BEGINS	
Explore	Identify a priority to change
	Systematically explore programmes or practices to implement
	Look at how it will fit within school context
ADOPTION DECISION	
Prepare	Develop a clear plan
	Assess readiness of the school to deliver the plan
	Practically prepare to implement
DELIVERY BEGINS	
Deliver	Support staff and solve problems
	Reinforce initial training with support
	Use data to ensure faithful adoption and adaption
STABLE USE OF APPROACH	
Sustain	Have a plan for sustaining and scaling the plan from the start
	Keep acknowledging support and reward good practice
	Treat scaling up as a new process

Often, at the 'explore' phase, we might consider our moral purpose: that feeling we often get in our gut that something just isn't quite right. At that point, we may begin to explore the problem(s) in our classroom and classrooms across schools to prepare a plan. This could be about a focus on a coherent curriculum, the use of representations to reveal mathematical structure, embedding mathematical vocabulary in lessons or something else.

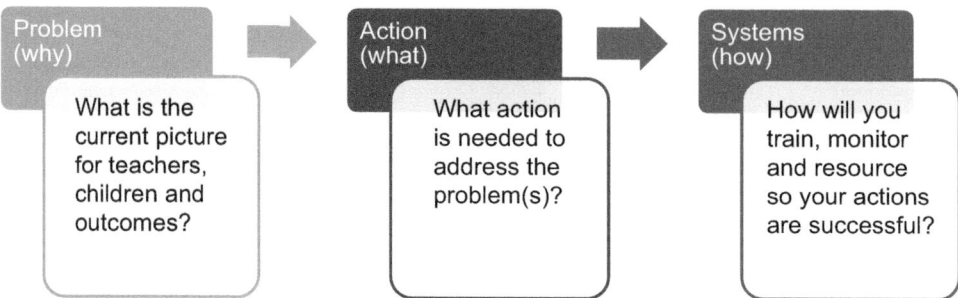

Problem (why)
What is the current picture for teachers, children and outcomes?

Action (what)
What action is needed to address the problem(s)?

Systems (how)
How will you train, monitor and resource so your actions are successful?

FIGURE 1.4 Start with your why improvement plan

Consideration should be given as to how the plans feed into your long-term goals and what outcomes for staff and children will look like in the short, medium and long term. Many schools do this in reverse and start with *what* they will do, rather than asking *why* there may be an issue. Starting with the big questions of 'why' can be useful before considering your 'how' or 'what', and will focus your thinking on an element of improvement.

From my experience, involvement from senior leaders is essential to this. Starting the conversations with your 'why' – what is currently happening and the change needed – is your rationale for change with moral purpose. It makes conversations with staff teams and all stakeholders very clear as to what your vision is and the steps that you will take to get there.

Systems

The decisions that we make as teachers or leaders should engineer success. Systems of high-quality professional development should lead high-quality teaching, and in turn lead us to require fewer interventions so that children can keep up with the intended curriculum. As we will consider in later chapters of this book, curriculum construction, sequencing and improvement are important systems to get right. Developing teachers with strong subject knowledge and pedagogical knowledge takes time, but the systems that we use for this can drive improvement.

I am sure that many of you reading this have adopted a scheme or approach. In many ways, this is the start of building strong systems for maths. All schemes and resources can only be as good as the person delivering them. The implementation process will consider the small component parts to develop across the school.

EYFS

All children will arrive in schools in nursery or Reception with varying degrees of exposure to mathematics, just as all children arrive at different attainments each year after the summer holidays. Some children may have sung mathematical songs (a personal favourite of mine is 'Five little speckled frogs sat on a speckled log', which reveals the structure of subtraction as taking away). Some children may know which house number they live in or they might have seen numbers on buses, on football shirts or in supermarkets. Some may have played board games and be experienced with turn-taking, identifying the cardinality of number by using dice and using mathematical vocabulary. Others may not have done any of these things.

When children start school in the Early Years, plans should consider the school entry gap in terms of knowledge of number. Children may be able to count, and indeed some parents may insist that their child can count to 100 and needs to be moved on, but this misses the point. While counting is important, so is understanding the ordinal (positional) value, cardinal ('how manyness') value and composition (numbers made of two or more smaller numbers) of numbers. Understanding their structure should be deliberately and systematically planned for. Having key number facts identified may be the starting point, and consideration as to how and when these will be taught is time well spent. The aim here is to build strong foundations on which children can build as they enter Key Stage 1.

Key takeaways

In conclusion, this introduction has laid the groundwork for understanding the principles of teaching for mastery in maths. We have explored the transformative shift from traditional methods, which often relied on rote memorisation and superficial understanding, to a more holistic approach that emphasises deep, conceptual learning. The five big ideas – coherence, representation and structure, mathematical thinking, fluency and variation – serve as the pillars of this educational philosophy. Teaching for mastery involves ensuring that all students, regardless of their perceived abilities, achieve a profound and connected understanding of mathematical concepts. This approach rejects the practice of ability grouping, which often limits students' potential, and instead promotes high expectations for all.

Moreover, we have discussed the importance of retrieval practice and the concept of 'stage not age', highlighting that learning should be tailored to each student's current understanding rather than their chronological age. By focusing on foundational concepts and ensuring that these are mastered before moving on, we can prevent the gaps in knowledge that have historically plagued maths education.

As subsequent chapters will explore, the crucial role of specific and technical maths vocabulary is at the heart of this book. I will advocate for precise and consistent use of mathematical terms to enhance understanding and communication. As we move forward in this book, each chapter will delve into these principles in greater detail, providing practical strategies and examples to implement teaching for mastery effectively.

Ultimately, the goal is to create a classroom environment in which every child is given the opportunity to succeed and develop a lasting, meaningful understanding of maths. Through collaboration, professional development and a relentless focus on equity, we can transform our teaching practices and ensure that all students are equipped with the skills and confidence that they need to excel in maths and beyond.

2 Coherence

Mathematics makes sense to mathematicians because they have a developed sense of the components and concepts of maths and how they all fit together into the bigger picture. Ideas flow and develop logically. Children (and teachers) who rely on tricks, short-cuts and gimmicks (superficial explanations or methods) will only get so far in primary school before ideas and concepts require a deeper understanding of the structure of mathematics in order to solve problems effectively. For example, I am sure that many of us have taught the 'bus stop' method for short division. The same is true for crocodiles when comparing numbers using the inequality symbols (<, > and =) to show children that the crocodile eats the bigger number or, when rounding, teaching children '1, 2, 3, 4, go to the number on the number line before, 5, 6, 7, 8, 9, go to the next 10 on the number line' – a complicated rhyme that underdevelops the real structure at play here (not to mention the fact that it may not be the nearest 10 we are rounding to and there may or may not be a number line used). It is therefore not difficult to see how these gimmicks offer no or little depth of understanding, and so may limit the flow of thinking at later stages of education. This is true for those children who are seen as 'good at maths'. They may be able to do lots of different things, but each time they do them, they are still having to retrieve the short-cut or gimmick in order to answer successfully, rather than focusing on the underlying structure. In doing this, mathematics becomes an over-elaborate memory game. At some point, as new ideas need to be learned and assimilated, that memory will begin to waver, and that's when maths becomes difficult for children. Moreover, this may be one of the reasons why adults openly admit that they are 'not good at maths'. The way in which maths is taught, the pedagogy behind it and how ideas are represented to connect with each other are very important to ensure high levels of success.

If the way in which a concept is taught is different in each year group, misconceptions may develop. Knowing the structure of the number system can be taught using number lines from Year 1 (starting in the Early Years Foundation Stage) to Key Stage 2 and beyond.

Teaching for depth

What does teaching for depth of understanding look like in reality? Let's think about the concept of rounding and the structures that sit behind it. How do the ideas connect to each other? And how does this concept develop across the primary age phase? Rounding is concerned with the closeness or proximity of numbers. This learning starts in Year 1, where children reason about the location of numbers on linear number lines and begin to compare numbers using the inequality symbols. In Year 2, we may expect children to reason about the location of any two-digit number on a linear number line and identify the previous and next multiples of 10 of given numbers. In Year 3, children then focus on the location of any three-digit number in the linear number system and identify the previous and next multiples of 10 and 100. By Year 4, children have had lots of practice at finding where numbers go and previous multiples, and so the concept of rounding to the nearest 10, 100 and/or 1,000 may be introduced, alongside the learning that children have had on place value and four-digit numbers. In Year 5, this concept is extended to reasoning about the location of numbers with up to two decimal places in the linear number system and identifying the previous and next multiples of 1 or 0.1. By Year 6, children may then reason about the location of any number up to 10,000,000 and round to any given context. And to think – this concept started developing in Year 1, with learning where numbers are placed on a number line.

Conceptual progression

This idea of conceptual progression allows children (and teachers) to see where a concept begins and how a concept develops. By the end of Year 6, once children have mastered this concept, they are ready to start Year 7 and to deepen and strengthen it, including rounding digits to a required number of decimal places, rounding numbers to a required number of significant figures and estimating calculations by rounding (including contexts of measure).

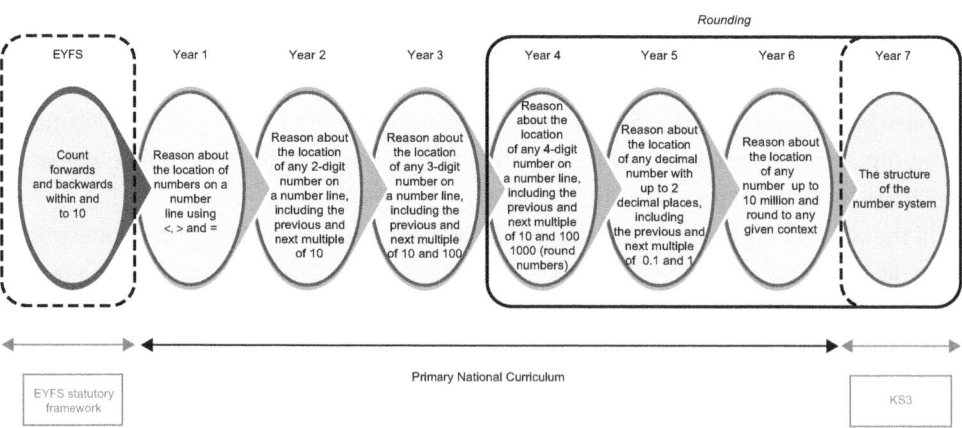

FIGURE 2.1 Progression in a concept from Early Years to Key Stage 2 and beyond

Knowing how ideas are connected, where they start and how they develop is essential to effective teaching and learning if we want children to master the underlying structures that are at play. The number lines shown in Figure 2.2 demonstrate how a solid understanding of the structure of the number system can support children when they encounter rounding in Year 4 and beyond, based on the 'closeness' and proximity of numbers in a given context. This way of teaching number lines is effective for helping children to master the underlying structures at play, because it means that they understand how the numbers are connected, where they start and how the number line develops, without the need to rely on tricks, short-cuts or gimmicks.

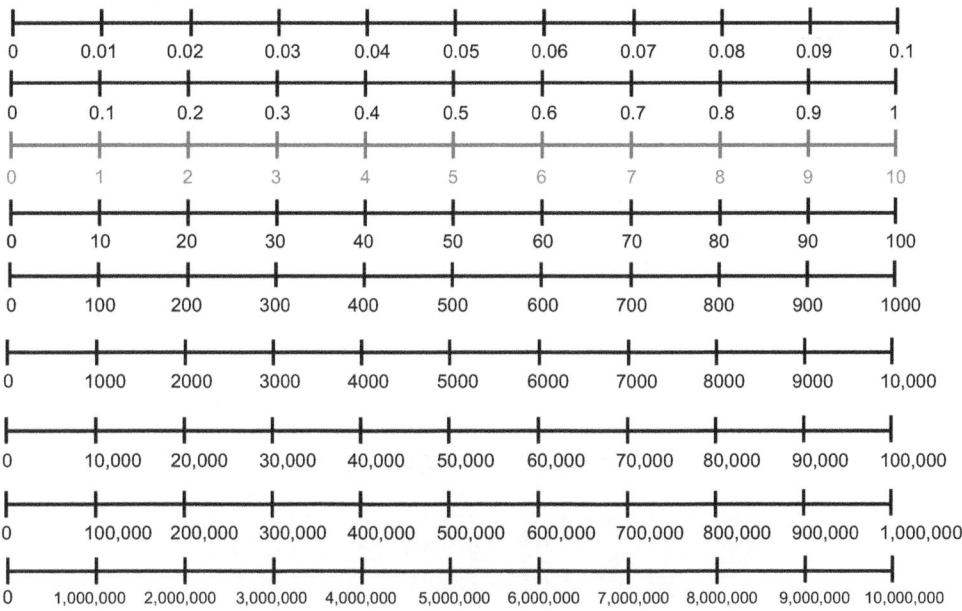

FIGURE 2.2 Number lines revealing the structure of the number system

This coherent and systematic way of teaching the structure of the number system from EYFS allows us to make links to other areas of the curriculum, such as comparing numbers and rounding. We can see how if children know 3 < 7, then we can later teach 0.3 < 0.7 and therefore 0.03 < 0.07 and, of course, 0.3 < 0.05 + 0.02, for example. We can design much of the curriculum with the number line in mind, using coherent, logical steps.

Let's now turn our attention to geometry and explore how representations reveal the structure of the maths being taught.

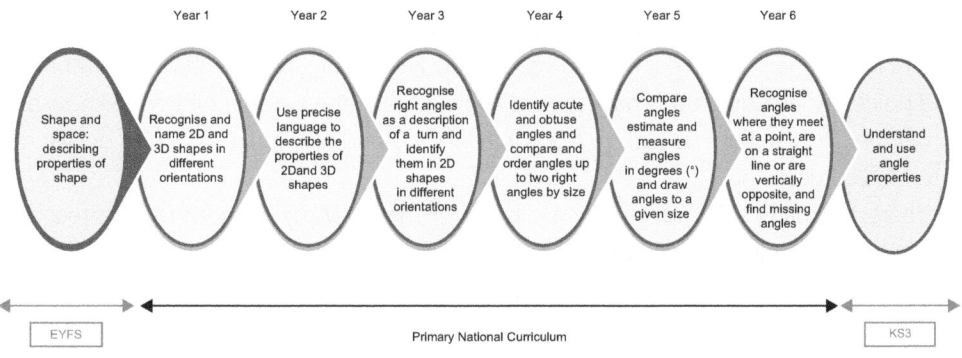

FIGURE 2.3 Progression in the concept of geometry from Early Years to Key Stage 2 and beyond

The image in Figure 2.3 demonstrates what coherence can look like in mathematics. Concepts that build on prior knowledge can allow children to reason and apply their understanding in a range of contexts. However, we can think of coherence in a different way. We may want to consider what a coherent curriculum looks like – a coherent unit of work, a coherent mathematical year and, at the macro size, a coherent curriculum. This requires a great depth of subject knowledge of how all of the pieces or concepts fit together, how they can be developed over time and where they link to other areas of the curriculum. For example, there is a reason why place value, the four operations and fractions tend to be taught towards the beginning of the academic year – it is because these concepts are fundamental to many areas of mathematics. Learning to count in multiples of 5, 10, 25 or 100 may pay off in dividends when children are interpreting graphs and charts with such scales; learning to calculate with fractions may support understanding of area and/ or perimeter (a triangle being half the area of a rectangle); and recall of times tables facts is vital in order to simplify fractions. They all link. They are all connected.

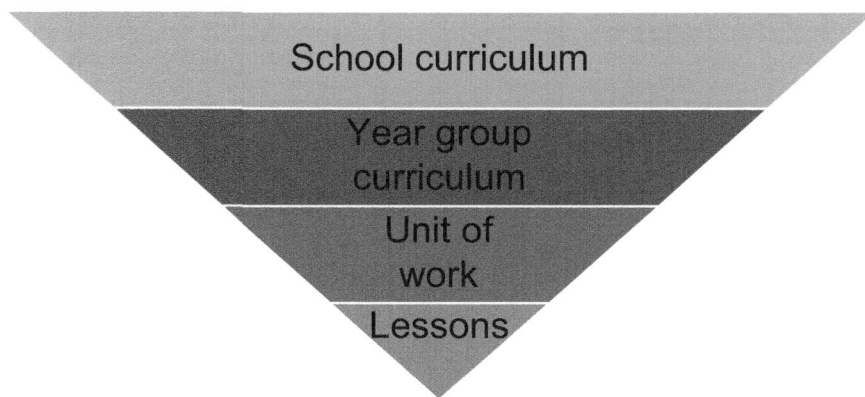

FIGURE 2.4 Whole-school thinking and leadership impact on day-to-day lessons

Whole-school curriculum

We should first start with a whole-school curriculum: an overview of the concepts to be taught across the school year in each year group. From there, we can chart our ambitious end points by including the small steps of knowledge that we want children to know and remember. We can then use this to plan units of work and individual lessons to deliver our intended curriculum.

There are many schemes readily available for maths. Many of them require teachers to log in to download lesson presentations, with some suggested activities for children to do. There is a danger that this could de-skill teachers, as it limits thought around *how* concepts may be taught. It may also limit the thinking around which representations are best used to support understanding, which vocabulary could be introduced or used, what reasoning might look like in the lessons and across lessons, in which elements of the curriculum children need to be fluent and opportunities for variation theory and intelligent practice. Over the years, I have seen too many false starts with curriculum, and this is where it must start. Even with readily available schemes, careful professional development for all colleagues, teachers and teaching assistants is essential.

Teaching in blocks

Some schemes 'block' units across the year. For example, geometry or statistics may be taught in a block each year. While this seems intuitive in terms of sequencing learning within these units in small steps, we ought to consider what this means in terms of children actually knowing and remembering more. We could formally teach the block and move on, but at what expense? With any age group I have taught, I have always had to go back to very early objectives, such as naming 2D and 3D shapes and their properties, because children have forgotten them after not revising them regularly within the curriculum. Perhaps revisiting, reteaching and reminding children regularly about these units needs to be deliberately planned and thought out. Option 1 shows a curriculum designed using blocking; this can promote deep learning and allows concepts to breathe and small steps to build logically and link lessons together. Option 2 shows a blocking approach with deliberate and systematic retrieval built in and Option 3 shows a shorter version of this.

Option 1

TABLE 2.1 An option for blocking units

Autumn	Place value	Addition and subtraction	Area	Multiplication and division		
Spring	Multiplication and division	Length and perimeter	Fractions	Decimals		
Summer	Decimals	Money	Time	Shape	Statistics	Position and direction

Option 2

This option prioritises shape, as this can often be an area of the curriculum seldom revisited in any depth throughout the year.

TABLE 2.2 An option for blocking units with built-in retrieval

Autumn	Shape from previous year group(s)	Place value	Shape from previous year group(s)	Addition and subtraction	Shape from previous year group(s)	Multiplication and division					
Spring	Shape from previous year group(s)	Multiplication and division	Shape from previous year group(s)	Length and perimeter	Shape from previous year group(s)	Fractions	Shape from previous year group(s)	Decimals			
Summer	Shape from previous year group(s)	Decimals	Shape from previous year group(s)	Money	Shape from previous year group(s)	Time	Shape from previous year group(s)	Shape	Shape from' previous year	Statistics	Shape from this year	Position and direction

We might even consider a more spiral approach and prioritise the number elements of the curriculum by spacing them across the year so that children are constantly revisiting key ideas.

Option 3

Here we have shorter units, with scope for the previous year's (or years') concepts being taught before introducing new learning. This can allow teachers to 'track back' to ensure that children build on strong foundations. Of course, this may mean lesson time, and perhaps less depth, being spent on some topics.

There are advantages and disadvantages to each model. Whichever you choose, having a clear rationale about which staff, children, governors, parents and all stakeholders are clear is essential. Start with why you want to design your curriculum in a particular way and then build from there.

TABLE 2.3 An option for shorter blocking of units with built-in retrieval

	2/3 weeks		2/3 weeks		2/3 weeks		2/3 weeks		2/3 weeks		2/3 weeks
Autumn	Retrieval of learning from previous year	Place value	Retrieval of previous addition and subtraction	Addition and subtraction	Retrieval of previous year shape	Shape	Retrieval of place value from this year	Place value	Retrieval of addition and subtraction from this year	Addition and subtraction	
Spring	Retrieval of previous year objective	Multiplication and division	Retrieval of multiplying (ready for area)	Measure (area)	Retrieval of previous learning from this year	Place value	Retrieval of previous year objective	Fractions	Retrieval of shape objectives	Length and perimeter	Statistics
Summer	Retrieval of previous learning from this year	Fractions	Retrieval of previous learning from this year	Multiplication and division	Retrieval of place value from this year	Decimals	Retrieval of multiplication and counting in 2s, 5s and 10s (ready for money)	Money	Retrieval of previous learning from this year	Decimals	Position and direction

S-planning

S-planning can be an effective way in which to analyse the structure of a curriculum, whether it be a readily available one, one that a school, trust or local authority has developed or a curriculum that is cherry-picked and bespoke. All of these models are sound if they are implemented effectively. Knowing your colleagues, looking at what the data suggests and knowledge of your school are important when choosing what action(s) to take. Let's suppose that you want to look at where connections can be made across the curriculum. An S-plan offers a lens through which to look at the curriculum and where the connections can be made. This is the time to get out a large piece of paper and colourful pens and begin the creative process of curriculum design.

In the example in Figure 2.5, a typical Year 5 sequence is shown. This is the starting point for any teacher. Your Year 5 teacher(s) may work on this in a staff meeting alongside the curriculum that you use. Through the lens of coherence, you might ask how each of the units link to each other. For example, at which points will multiplication and division be used across your curriculum? What does it look like? How does multiplication in Year 5 link with the teaching of fractions? Or area and perimeter? Or decimals and percentages? Similarly, you may ask how shape links to areas of the curriculum. For example, how does understanding of rectilinear shapes link with area and perimeter? Moreover, what are the links to multiplication and division? Or how does understanding of shape link to fractions of a shape or decimals? Then, how does understanding of both multiplication and shape link with teaching volume? This will bring clarity to thinking around how all of the ideas are connected and flow into each other, rather than being unrelated, discrete units taught in isolation. Once this activity is carried out with colleagues, you could team people up with adjacent year groups to discuss the links and to see where a concept is taught in the previous year groups(s) and where the concept develops in later year groups. This way of working on your curriculum design often results in very colourful displays of how your curriculum is implemented and how all of the concepts link together.

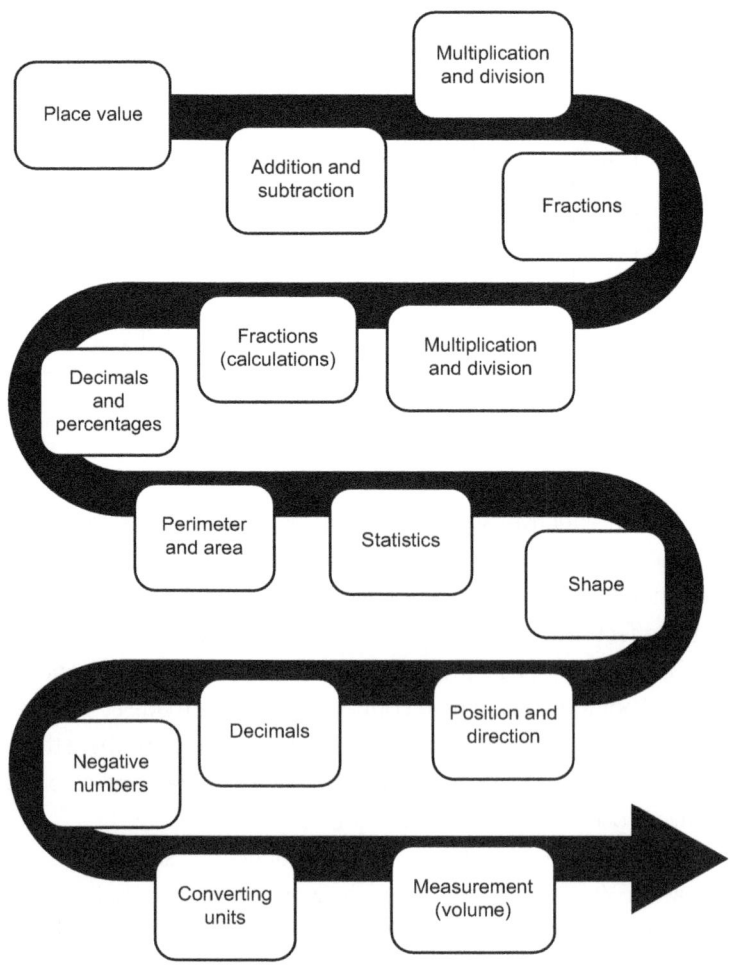

FIGURE 2.5 An S-plan showing the sequence of units in a year group

S-planning can be done in so many different ways. It can link directly to your action plan or school improvement plan. You could revisit the S-plans to start across a term or across the school year, but look at a particular focus each time (perhaps tracking the small steps for a fraction unit) and see how the ideas link together. It could be that you ask colleagues to add the specific vocabulary introduced or needed to understand a topic, which representations would be useful (get colleagues drawing them on their S-plan), which stem sentences could be used, which activities children would do and so on. Focus on one thing at a time, and over time colleagues will build a rich and deep network for your curriculum, how it works and how ideas fit together. Most importantly, they will deliver lessons that make sense to children.

It's clear to see that coherence is the glue that holds curriculum and pedagogy together. To learn something new, a new concept is connected or incorporated into the learner's existing knowledge (assimilation) or existing knowledge is extended or reorganised (accommodation) (Skemp, 1986). We can see how S-planning may go some way to support teachers' and teaching assistants' understanding of curriculum, how knowledge is connected in the year group that they teach and how a concept develops from previous year groups. Essentially, this is what we mean by small steps: breaking learning down into small, manageable chunks that allow children to practise and repractise a particular skill. As educators, it is our duty to ensure that we know how that skill connects to the bigger picture in the next lesson, how it connects in the next week or unit of work, how it connects across the whole year group curriculum and, ultimately, where it sits across the whole primary curriculum. Focusing on one key point in a lesson allows for deep and sustainable understanding. Only when something has been mastered can it be used to build new learning and create new schemas.

An S-plan can act like a roadmap or satnav, but even with the sharpest attention, you can still end up down some roads that you don't need to be down. This is where our teacher knowledge is necessary. Knowing which turn to take and when is important; knowing when to take in the scenery or when to turn around and re-route are all effective strategies. Assessment for learning (AFL) is pivotal to this approach. Imagine that you are going on a holiday abroad and to get there you need to take a flight. The moment at which you arrive at the airport, check in, go to departures, board the aircraft, take off, order a drink and food, read, debate whether to go to the toilet, wonder why there is a strange noise coming from the engine, wonder why the wings seem to be wobbling, wonder when the screws were last tightened on the wings and then hear 'Ten minutes until landing' before touching down on the tarmac – at each stage of your journey, you are supported and helped by staff. Your comfort and every need are met at every available opportunity, and adjustments are made where they can be. Teaching works a little bit like this. We wouldn't embark on the flight of a lesson without systematically checking the comfort of our passengers (the children) in front of us, particularly when our children are novice flyers and some, of course, may well not have ever been on an aeroplane before.

Mixed-age classrooms

This section of thinking is from my work in mixed-age settings. However, many of the underlying principles apply to effective teaching in any classroom. In distinct year groups there are mixed attainments, just as there will inevitably be in mixed-age classrooms.

The National Curriculum is designed in clear year groups, with associated end-of-year expectations. This requires careful thinking around how to successfully implement a curriculum in mixed-age classrooms. In some schools, classes may be split into Years 1 and 2, Years 3 and 4 and Years 5 and 6. Other examples could be EYFS and Year 1, Years 1 and 2, Years 2 and 3 and Years 3, 4, 5 and 6 all in one class. I have come across three main approaches, each with merit and limitations: split delivery, teach together and align and refine.

Split delivery

Split delivery is just that. The children in different year groups are taught separately. This could mean that the teacher works with Year 1 while a teaching assistant works with Year 2. This ensures that children from each year group get their curriculum entitlement and coverage, and single-aged schemes of learning can be used. However, we should be mindful that all children need daily teaching from a teacher, children need to be taught to work independently and, in some cases, the lesson can become confusing, due to two completely different lessons being taught – talk about cognitive overload!

Teach together

Imagine a mixed Year 5 and 6 class. They are taught together and for each area of maths, the teacher carefully assesses the starting point of the class. The class is taught a mixed Year 5 and 6 curriculum. This often means starting with Year 5 objectives and then moving on to Year 6 objectives later in the year. In doing so, the teacher is responsive to assessment; teaching is focused towards stages of development and not age; there is a clear progression in the learning sequence; and older children can recap and consolidate prior learning, not to mention different curricula for these year groups. Year 5 do not learn algebra or long division in the National Curriculum, so this can quickly become difficult and lead to split delivery. However, there is a lot of content to cover, and this can limit depth of understanding and lead to a coverage model. Older children may get bored by repeating content that they have been previously taught for whole lessons or weeks at a time, and the younger children may not (and should not) move on to the next year group, as this becomes 'a race to the top'.

Align and refine

This approach aligns similar objectives so that teachers can teach one concept and teach it well. Imagine a mixed Year 3 and 4 class. All children begin the lesson being taught the Year 3 content. Year 3 then access some independent work around what has just been taught. Year 4 children are then taught while the Year 3 children practise, to introduce new learning around a similar objective. This means that objectives ensure a similar focus for both year groups. It means that teaching is clear; concepts may be taught well; there are still opportunities for older children to revisit previously taught concepts and build; and there is clear progression between the year groups. This approach to curriculum does require careful thought about how to map out objectives and how assessment will inform planning and lessons. (As in the following example, teachers may need to track back to objectives prior to Year 3 for some classes.) I have seen this approach done most effectively. Table 2.4 shows how one strand of place value builds in each year group and thus allows teachers to 'scale' back.

TABLE 2.4 One strand of place value and how it builds in each year group (adapted from DfE, 2020) .

EYFS	Year 1	Year 2	Year 3	Year 4	Year 5	Year 6	Key Stage 3
Develop number sense by verbally counting forwards to and beyond 20, pausing at each multiple of 10	Count forwards and backwards within 100	10 ones are equivalent to 1 ten 10 tens are equivalent to 1 hundred	10 hundreds are equivalent to 1 thousand	10 one thousands are equivalent to ten thousand	10 tenths are equivalent to 1 one 100 hundredths are equivalent to 1 one 10 hundredths are equivalent to 1 tenth	All relationships with powers of 10	Use place value for decimals, measures and integers of any size

Looking closely at Year 3 and 4, we can see how those objectives could align well, how all children in a Year 3 and 4 class could access the Year 3 content at the start of a lesson and how the Year 4 children would build upon this understanding part-way through the lesson.

Let's look at this another way – Table 2.5 shows a Year 5 and 6 example, highlighting the small steps and how they link. In this example, the place value objectives are merged with negative numbers to bring greater coherence to the curriculum design. You will notice that some objectives span across both year groups – this is because the older year group is consolidating that skill and there may not be any additional content to teach in the older year group (see Roman numerals, powers of 10, partition numbers to 1,000,000 and find the difference objectives). Still, retrieval and consolidation can be an effective part of curriculum design and help children to know more and remember more.

TABLE 2.5 The alignment between Years 5 and 6

Year 5	Year 6
Roman numerals to 1,000	
Numbers to 10,000	Numbers to 1,000,000 Numbers to 10,000,000
Numbers to 100,000	Read and write numbers to 10,000,000
Numbers to 1,000,000	
Read and write numbers to 1,000,000	
Powers of 10	
10/100/1,000/10,000/100,000 more or less than a number	
Partition numbers to 1,000,000	
Number line to 1,000,000	Number lines to 10,000,000
Compare and order numbers to 100,000	Compare and order any integers
Compare and order numbers to 1,000,000	
Round to the nearest 10, 100 or 1,000	Round any integer
Round within 100,000	
Round within 1,000,000	
Understand negative numbers	Negative numbers
Count through zero in 1s	
Count through zero in multiples	
Compare and order negative numbers	
Find the difference	

In some cases, more than one objective has been aligned. This is because it is a similar concept but there is scope within a lesson or series of lessons to go deeper within that objective, based on teacher assessment and how much practice children need.

Similarly, the Year 1 and 2 planning in Table 2.6 aligns objectives in place value and makes connections with money.

TABLE 2.6 The alignment between Years 1 and 2 using money

Year 1	Year 2
Count within 20	Count money (pence)
Understand 10	Count money in pounds (notes and coins)
Understand 11, 12 and 13	Count money (pounds and pence)
Understand 14, 15 and 16	Choose notes and coins
Understand 17, 18 and 19	Make the same amount
Understand 20 Compare numbers to 20 Order numbers to 20	Compare amounts of money
1 more, 1 less	Calculate with money
Number line to 20	Make a pound
Use a number line to 20 Estimate on a number line to 20	Find change
Consolidation of previous steps	Two-step problems

The planning in Table 2.7 shows how objectives are aligned with geometry (shape) in Years 1 and 2.

TABLE 2.7 The alignment between Years 1 and 2 using geometry

Year 1	Year 2
Recognise and name 3D shapes	Recognise 2D and 3D shapes
Sort 3D shapes	Count the faces on 3D shapes
Recognise and name 2D shapes	Draw 2D shapes
Sort 2D shapes	
Patterns with 2D and 3D shapes	Count the sides on a 2D shape Count vertices on a 2D shape
Consolidation based on assessment	Lines of symmetry and shapes Use lines of symmetry to complete shapes

I mentioned less common splits earlier; Tables 2.8 and 2.9 are examples for Years 4 and 5. This is curriculum planning to align fractions and then decimals.

TABLE 2.8 The alignment between Years 4 and 5 using fractions

Year 4	Year 5
Understand the whole	Find fractions equivalent to a unit fraction
Count beyond 1	Find fraction equivalent to a non-unit fraction
Partition a mixed number	
Number lines with mixed numbers	
Compare and order mixed numbers	Compare fractions less than 1 Order fractions less than 1 Compare and order fractions greater than 1
Understand improper fractions	
Convert mixed numbers and improper fractions	Convert mixed numbers to improper fractions
Convert improper fractions to mixed numbers	Convert improper fractions to mixed numbers
Equivalent fractions on a number line Equivalent fraction families	Recognise equivalent fractions
Add two or more fractions	Add and subtract fractions with the same denominator Add fractions within 1 Add fractions with a total greater than 1
Add fractions and mixed numbers	Add to a mixed number Add two mixed numbers
Subtract two fractions	Subtract fractions
Subtract from whole numbers	Consolidation of subtracting fractions
Subtract from mixed numbers	Subtract from a mixed number Subtract from a mixed number, breaking the whole Subtract two mixed numbers

TABLE 2.9 The alignment between Years 4 and 5 using decimals

Year 4 (Decimals)	Year 5 (Decimals and percentages)
Tenths as fractions	Thousandths as fractions
Tenths as decimals	Thousandths as decimals
Tenths on a place value chart	Equivalent fractions and decimals (tenths)
Tenths on a number line	Equivalent fractions and decimals
Divide one digit by 10	

Table 2.9 (Continued)

Year 4 (Decimals)	Year 5 (Decimals and percentages)
Divide two digits by 10	
Hundredths as fractions	Equivalent fractions and decimals (hundredths)
Hundredths as decimals	Decimals up to two decimal places
Hundredths on a place value chart	Thousandths on a place value chart
Divide a one- or two-digit number by 100	
Make a whole with tenths	
Make a whole with hundredths	
Partition decimals	
Flexibly partition decimals	
Compare decimals	Order and compare decimals (same number of decimal places)
Order decimals	Order and compare any decimals (up to three decimal places)
Round to the nearest whole number	Round to the nearest whole number Round to one decimal place
Halves and quarters as decimals	Understand percentages Percentages of fractions Percentages as decimals Equivalent fractions, decimals and percentages

Again, S-planning can be a useful tool with mixed-age classrooms and allows teachers to understand the structure of their curriculum and how ideas are linked and connected and flow into one another. There is no silver bullet, and even this model has limitations. I have found that some year group objectives do not align very well, and so the majority of teaching is aligned but there may be a few weeks of the year where split delivery of distinct objectives is needed.

The model in Figure 2.6 may be useful in supporting your thinking about the structure of a mixed-age lesson of Year 3 and 4 children. As we have discussed, the core objective would be identified and aligned with the previous objective from Year 2. This may be the starting point of the lesson – to recap the concept being taught for a few minutes. The main input would be focused around the Year 3 content, while this is still valuable retrieval for Year 4. During the tasks, both Years 3 and 4 may work on them but, at some point, a decision should be made on when to change this and teach the Year 4 children the same concept with the aligned objectives from Year 4. In most cases, this will build on what has already been taught at the start of the lesson. This way, all children receive high-quality teaching.

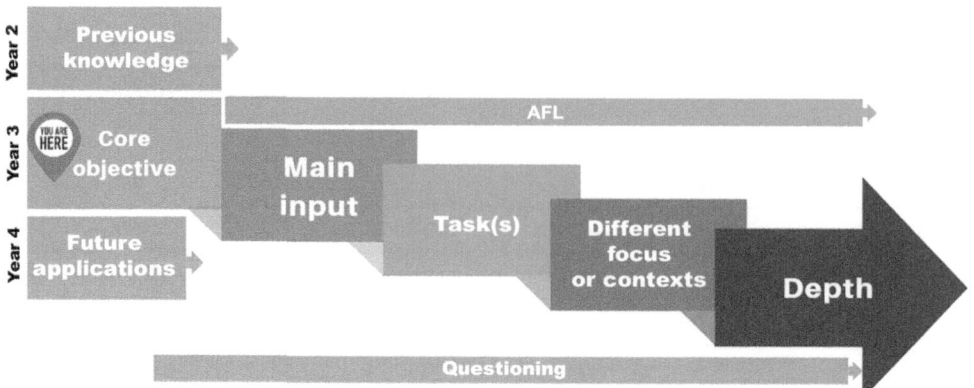

FIGURE 2.6 A suggested model for teaching mixed-age classes

The National Curriculum for mathematics is a cumulative curriculum. Each year builds on previously taught concepts. Aligning curriculum objectives allows teachers to teach lessons that spiral and uses retrieval as a key approach to reteach and revisit previously taught concepts. The power of the approach relies on AFL in lessons and knowing when and, most importantly, how to ensure that children gain a depth of understanding.

Small steps and lesson design

We have alluded to small steps and their importance in coherent lessons. I'd now like to take you through the process of planning a lesson, including small steps. It turns out that the small steps have even smaller steps. Published schemes have some suggestions on how lessons could be structured, but it is important to consider the pedagogy that sits behind the planning. I am going to design this lesson with you. It is a Year 6 lesson, but these principles can be applied to any lesson. Imagine that I have downloaded some planning and a lesson presentation, but I want to deliver this lesson in an exciting way that goes beyond using the presentation. This lesson context is on calculating the area of any triangle. The children have already counted squares and calculated the area and perimeter of rectangles, and AFL tells me that my class are ready to move on.

Shapes – same area	Area and perimeter	Area of a triangle (counting squares)	Area of a right-angled triangle	Area of any triangle	Area of a parallelogram	Volume – counting cubes	Volume of a cuboid

What do I want the children to know by the end of the lesson?

I want the children to know and generalise that a triangle is half the area of a rectangle. I want to lead them towards the generalisation that:

area of triangle = ½ x base x height

I'm not a huge fan of sharing an intended learning objective every lesson. For me, it's like giving away the punchline to a joke. A few years ago, there used to be a TV show in the UK called *Take Me Out* and it got me thinking about lesson design (stick with me on this one). During the show, 20 or more women would have their light switched on to show their interest in a bachelor who arrived in the 'love lift' to present themselves and offer up a date. This person would say a couple of sentences about themselves and the women would switch off their lights if they were not interested. For me, lesson objectives can work in a similar way. If we share something at the start of a lesson – 'Today we are going to be learning about the area of triangles' – and children don't really want to do it, we can often see the lights switching off behind their eyes before the lesson has even started, or they have fixed mindsets on area or geometry.

So, instead, to start this lesson, I give children some coloured card and ask them to 'follow me' while I draw some rectangles under the visualiser and demonstrate on the screen (a great opportunity to measure accurately). I've thought carefully about the example of the rectangles that I want to show children. One rectangle is 2 cm x 5 cm and the other is 2 cm x 10 cm (I'll make links to this later).

Once these have been drawn accurately, I ask children to cut them out and ask them what they have noticed. After visiting some groups of children, I draw together responses. I then overlay the two rectangles and begin to focus their thinking on the area of each shape. I ask children to use the words 'double' or 'half' with area in their explanation, and perhaps write them on the working wall next to the rectangles that I have drawn. I am drawing attention to 2 cm x 10 cm = 20 cm² and 2 cm x 5 cm = 10 cm² being doubles and halves.

Notice the 'ping pong' approach to this teaching – systematically checking pupils' comfort throughout the 'flight' of the lesson. Some schools adopt an 'I do, we do, you do' approach so that teaching is explicit and direct. This can help lessons to be broken down into small coherent steps and get teachers thinking about what they will be doing and what the children will be doing at each stage of the lesson. In doing so, learning builds across the lesson (and across lessons) and teachers can adapt their teaching, spending longer on certain points if needed, based on what children have learned and can do.

I then ask children to 'follow me' again while I show them how to halve both rectangles diagonally to reveal a triangle. Without measuring, I ask what they now know about the triangles. I focus them on the area being half of what it was originally as a rectangle. I'm not yet ready to reveal the 'punchline' or key learning of the lesson. I want the children to really think deeply.

I then ask the children to 'follow me' under the visualiser again. This time, we draw two different triangles. I introduce the vocabulary 'base' and 'height' and ask children to draw one triangle with a 3 cm base and 6 cm height and another with a 6 cm base and 12 cm height. Again, I ask what they notice and draw out the relationship between the numbers. I then ask pupils what the area of the rectangle as which this triangle started its life as would be. This requires some creative thinking, and if children cannot visualise this, I show them by drawing out each rectangle and demonstrating that half of it makes the triangles that we have just drawn.

I now ask children to draw and cut out the following rectangles (but I'd ask them to draw out two of each):

- 2 cm x 4 cm
- 4 cm x 4 cm
- 8 cm x 4 cm

Then I ask children to calculate the area of these rectangles and think about what they notice. We explore the relationship between the measurements and the impact that they have on the area. The rectangles are cut up, stuck in books and annotated.

- 2 cm x 4 cm = 8 cm^2
- 4 cm x 4 cm = 16 cm^2
- 8 cm x 4 cm = 32 cm^2

Next to them, I ask children to visualise what the area of each of those measurements would be if they were triangles. Here, I'm going right in with the new learning. If children need some support, I show children that the triangle's area is half of the rectangle, as in Table 2.10.

TABLE 2.10 The relationship between the area of rectangles and the area of triangles

Area of rectangle	Area of triangle
2 cm x 4 cm = 8 cm^2	½ x 2 cm x 4 cm = 4 cm^2
4 cm x 4 cm = 16 cm^2	½ x 4 cm x 4 cm = 8 cm^2
8 cm x 4 cm = 32 cm^2	½ x 8 cm x 4 cm = 16 cm^2

Once children have completed this and they have halved the rectangles and shown the relationship in their books, I ask children to generalise this rule and teach them that the area of a triangle = ½ x base x height. I write this on a long strip of card and place it on the working wall next to some rectangles and matching triangles, to remind children of the rule in the lessons ahead.

To finish this lesson, I provide children with some carefully chosen examples of calculating the area of triangles by applying the rule that they have just explored. I ensure that the examples are carefully chosen so that children can look for patterns and links between the triangles. I provide coloured card, pencils and rulers to explore children's thinking *should* they need more practice with seeing the rule. At this stage, they are all right-angled triangles, as the next lesson would focus on perpendicular height and special cases. In the next lesson, I would start by revising and (more than likely) asking children to draw some rectangles and then calculating their area and the area of the triangle from that rectangle. I may also have a table (like the one in Table 2.10) with some missing calculations on each side so that children are looking for the connections and links. To go deeper, I may ask children to draw a 5 cm x 10 cm right-angled triangle accurately in their book and explain why the area of this is 25 cm² rather than 50 cm². Again, coloured card, pencils and colouring pencils would be available so that children could create and explain their thinking.

The design of these lessons moves them beyond a presentation and allows children to experience and behave as mathematicians. Teaching is direct and clear and scaffolds understanding with plenty of opportunities for purposeful practice woven throughout the lesson.

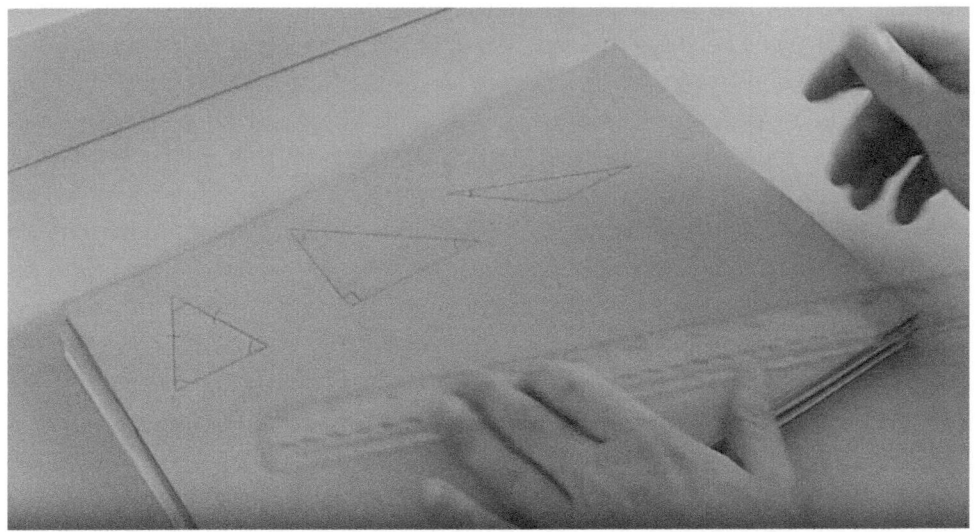

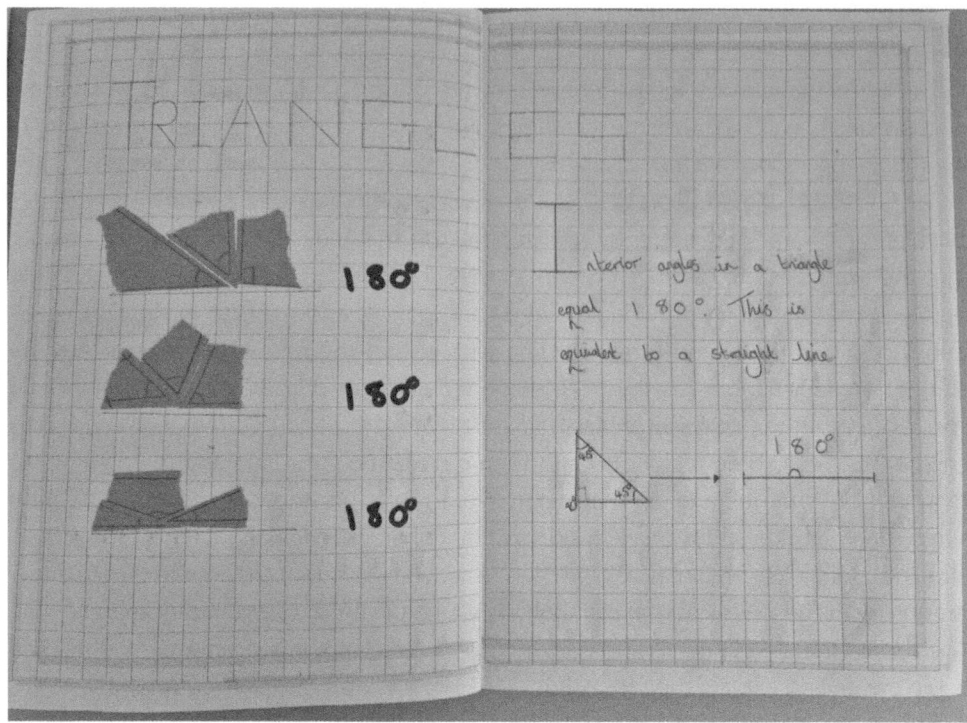

FIGURE 2.7 A maths book showing the connection between interior angles in a triangle and straight lines

And to think – I could have delivered a presentation and given pupils a worksheet, but the lesson described here allows for rich curriculum experience in maths. The small steps and practice that children have had and the way in which the lesson was designed allow children to build upon what they already know and to assimilate new knowledge. Imagine the power of beginning the next lesson by opening a child's book and asking them to explain the relationship between rectangles and triangles. This is retrieval at its most effective, built with small, coherent steps. This new knowledge in the form of a generalisation will be useful when calculating the area of any triangle in subsequent lessons. Depending on your class, you may look further into the components of this concept and perhaps focus on times tables at the start of the lesson, so that pupils' recall is quick and reliable in readiness to apply to the concept of area.

We can apply this thinking, using the five big ideas, to many areas of the maths curriculum. Planning small steps and going through the process of designing your teaching to break down the structure of the maths that you are teaching, alongside the mathematical representations and mathematical thinking (including which key vocabulary you want to teach children), is a good starting point. We should then consider what we want children to notice and what they will do to practise noticing this.

Early years

Our curriculum starts in EYFS. It should build the foundations for the whole-school curriculum. Each half-term should set out the specific knowledge, skills and vocabulary that children should be taught, and deliberate choices should be made as to the rhymes, songs and manipulatives that will be modelled to children. The messaging for Early Years is no different. We want children to understand concepts and ideas in depth, prioritise key areas of the Early Years curriculum (number) and ensure that children can subitise. This is where mastery begins – the belief that all children can achieve in maths and ensuring that children work together and progress through the curriculum at broadly the same rate. In these years, children are starting to form their self-image and beliefs around themselves as mathematicians.

There are six key areas of maths in the Early Years, three of which focus on numbers:

Number	Counting	Comparison	Composition
	Patterns	Shape and space	Measures

Counting to and from 20 is an important objective in the Early Years. The digits that we show children have a cardinal value, or the number of things that the digit represents. For example, the 'fiveness' of 5 is an important teaching point so that children understand what 6 can look like. In the Early Years, we may teach this using finger gnosis (recognising that fingers can represent objects) and/or a fives frame, tens frame or dice patterns (Hungarian number squares).

Comparison involves children knowing which numbers are more, less or equal to another number. Number lines are a useful representation to show the number system and for supporting counting forwards and backwards. Once this is secure, children can compare the relative size of numbers using the language of bigger or smaller.

Composition refers to children understanding that numbers can be made up of two or more numbers. For example, 5 is made of 4 and 1 or 3 and 2 (these are commutative facts) and 5 can also be made of 3 + 1 + 1 (also commutative) and so on. It is also linked to early understanding of addition and subtraction. Double-sided counters and fives or tens frames can be used to support number composition.

Effective teaching in the Early Years is led by adults. Learning cannot be expected to happen as children wander (with purpose) around a space, no matter how well put together that space is. Teaching should be daily and tightly focused around clear learning. The lesson should focus on one small idea or concept and be logically sequenced so that children can practise and rehearse the learning. Mastery should be achieved before moving on, with a strong focus on vocabulary. Direct modelling of manipulatives should be encouraged to reveal the structure of the maths that you are teaching. Throughout this, AFL should identify children who may require targeted support. Adults may guide enhancements in a provision. Of course, there should be opportunities for child-initiated learning. This is where your deliberate choices inform the thinking that children do. For example, if you have been teaching the composition of 5, it makes good sense to have this as enhanced provision so that children can choose to compose numbers to 5 to embed their thinking. The adults and environment should enable children to achieve well.

We should be clear about the difference between continuous provision and enhanced provision. Continuous provision is just as the name states: it is available continuously throughout the year. This could be a sandpit, water tray or construction area. Enhanced provision, however, links directly to what is currently or has been recently taught.

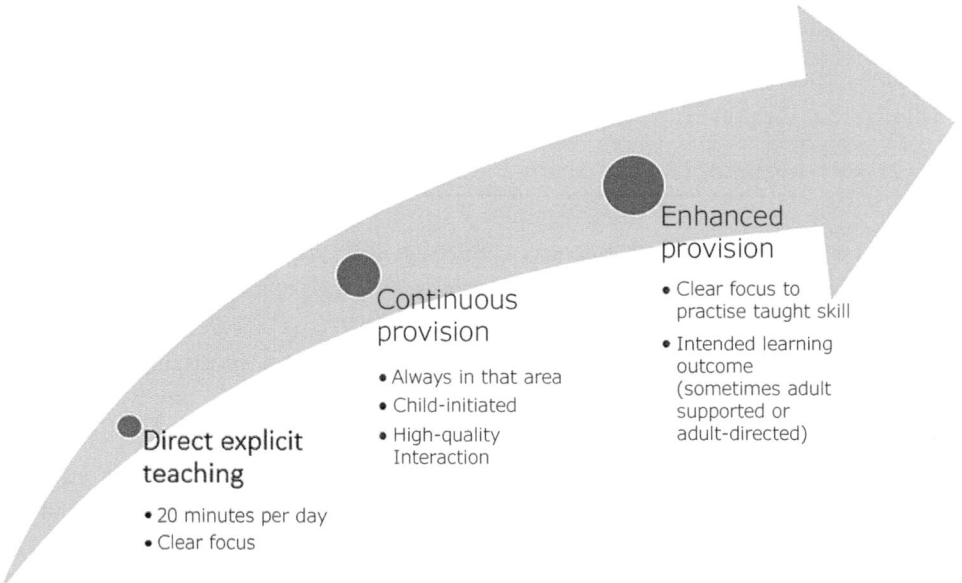

FIGURE 2.8 A suggested teaching sequence in Early Years

When monitoring in the Early Years, we should ask ourselves what the impact of our intended curriculum is and the rationale that sits behind the choices that have been made. We may ask about the strengths and areas to develop. These could be focused around adults' subject knowledge or pedagogical knowledge. When observing, we may focus on the quality of the interactions that adults have with children and how the provision supports teaching, as well as how all of this supports children in knowing more and remembering more.

Coherence

So, coherence is more than the small steps that we take. It involves how and why we design lessons in the way in which we do, the pedagogy that we use and the logical and clear way in which children progress through the curriculum. As we will see, this idea threads throughout all of the other ideas in this book, and works with the ideas rather than being separate. We will see how the idea of coherence can be applied to the representations that we use and how these reveal deep mathematical structures, and how we might design coherent tasks that allow children to make deep connections.

It is the cornerstone of a mastery approach, ensuring that each new concept builds logically upon the previous ones. This structured progression not only helps students to connect various mathematical ideas but also fosters a deep and enduring understanding. By employing strategies such as S-planning, educators can meticulously map out how concepts interlink across different units and year groups, reinforcing the curriculum's interconnected nature. This approach aids in tracking back to ensure solid foundations before introducing new content, thus supporting the development of robust mathematical schemas.

Key takeaways

Coherence in mathematics education involves recognising the importance of sequencing and structuring lessons in such a way that each new piece of knowledge is a natural extension of previous learning. This means designing the curriculum so that mathematical concepts and procedures are introduced in a logical order, allowing students to build on what they have already learned and to see the connections between different areas of mathematics. This cumulative approach helps to prevent gaps in knowledge and ensures that students have a solid foundation before moving on to more complex topics.

One practical application of coherence is the use of consistent representations and models throughout the teaching process. By consistently using visual aids, manipulatives and diagrams, teachers can help students to make sense of abstract concepts and see the connections between different mathematical ideas. For example, using number lines, bar models and part–whole models across various topics helps students to visualise and understand the relationships between numbers, operations and equations. This consistency in representation not only aids comprehension but also reinforces learning and retention. Be mindful of some 'off the peg' schemes that can overstimulate and confuse, with an overwhelming array of representations.

Another critical aspect of coherence is the careful planning of mathematical vocabulary. Ensuring that students understand and use precise mathematical language is essential for their ability to communicate their thinking and to understand the explanations of others. Teachers must introduce and consistently use the correct terminology, linking new vocabulary to previously learned terms and concepts. This practice helps students to develop a rich mathematical lexicon and enhances their ability to follow complex mathematical arguments.

Assessment for learning plays a vital role in maintaining coherence. By continuously assessing students' understanding, teachers can identify misconceptions and gaps in knowledge early and address them promptly. This formative assessment approach allows teachers to make informed decisions about when to review previous material and when to introduce new content, ensuring that each student progresses at an appropriate pace. Moreover, it enables personalised support, catering to the diverse needs of learners and ensuring that no student is left behind.

Ultimately, coherence is about seeing the bigger picture and ensuring that every small step contributes to a cohesive learning journey. It is the glue that holds the curriculum together, enabling students to build confidence and competence in mathematics. This coherent approach not only enhances students' ability to learn but also empowers teachers to deliver high-quality, connected lessons that make sense to all learners. By prioritising coherence through your curriculum design and lesson design, you can create a mathematics education that truly supports mastery, ensuring that all students have the opportunity to achieve deep and lasting mathematical understanding.

3 Representation and structure

The way in which we choose to represent mathematics is important, but the ultimate aim of mastering maths is that children do not over-rely on or need the representations at all, because the structure has been so well assimilated into their thinking. The NCETM states:

'The intention is to support pupils in "seeing" the mathematics, rather than using the representation as a tool to "do" the mathematics. These representations become mental images that students can use to think about mathematics, supporting them to achieve a deep understanding of mathematical structures and connections.' (NCETM, 2017)

Representations and structures are two separate things, but they aren't explicitly linked. The representations that we use in classrooms reveal mathematical structures, such as partitioning or additive and multiplicative structures. For example, a part–whole model is a representation, and it reveals the structures of parts and wholes of different mathematical concepts.

We should make the distinction between a representation and a manipulative. A manipulative is a physical tool that helps children to understand maths concepts – they are designed to represent and explicitly expose ideas that are abstract. An example of this might be base ten or tens frames (although there can be digital versions of these that children can move and interact with). Representations are visual or symbolic depictions of maths concepts. They show the *structures* that are being taught. For example, base ten may be used to show the structure of exchanging when adding or subtracting or a tens frame may be used to show number bonds to 10.

There are also different kinds of manipulatives: structured and unstructured. Structured manipulatives have been specifically designed to expose mathematical concepts. These are tools such as base ten or fraction strips. On the other hand, unstructured manipulatives are more open-ended and versatile and have not been designed with a specific mathematical purpose in mind. They can therefore be used in lots of different ways. Double-sided counters are an example of an unstructured manipulative that can be used in many different ways.

Consider this example. Which representations would you use to reveal the structure of the question?

John and Carl have some stickers in the ratio 1 to 4. Altogether, they have 50 stickers. If Carl gives half of his stickers to John, how many stickers will John have?

This problem involves comparing amounts, so we need a representation that would lend itself to comparison. An unstructured manipulative could be useful here. I've used this problem in Key Stage 2 and I've had success when using sticky notes to physically show children a comparison bar model. I've also used double-sided counters. We can start by representing the ratio and recognising that all of the parts or counters have a total of 50, and so each part has a value of 10.

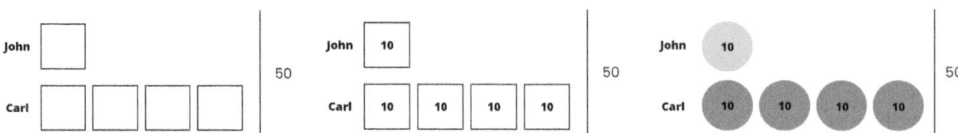

FIGURE 3.1 Representations reveal the structure of a problem

We can now see that half of Carl's amount is 20, and so if he gives it to John, John will have 30 in total. The manipulatives allow the teacher (or children, if you choose for them to use the manipulatives) to change the relationship. You could even use the question that we started with but change it slightly so that it keeps the same structure, and children will have plenty of purposeful practice with this. That's the power of using representations and manipulatives.

We may consider the big ideas of teaching for mastery together, but what would happen if we did not have representations to reveal mathematical structures in lessons?

Representation and structure	Fluency	Mathematical thinking	Coherence	Variation

We can look into the not-too-distant past, where maths teaching may have focused on unrelated rhymes, chants and gimmicks that do not show the maths that is working behind a concept. I'm going to put my cards on the table here and say that maths does not have anything to do with bus stops, crocodiles, putting numbers on doorsteps, adding zeros to numbers when multiplying by 10 or borrowing from numbers when adding or subtracting. Teaching in this way becomes an exercise in over-elaborate memorisation at the expense of really understanding the structures that are at work. Younger children may remember them – I am sure that there are adults out there who remember being taught in this way – but how confident can we be that they have actually understood the maths? As children progress from Key Stage 1 to Key Stage 2, and certainly beyond that, this way of teaching can slow progress, as children have to remember increasingly complex tricks and procedures. I'm sure that we've all taught children who confuse methods and try to retrieve the method that they have been taught – long division springs to mind. Before children can master this method, so many different components need to be secure so that children can draw upon them and make the connection between partitioning, multiples, factors, subtraction, remainders, decimals, rounding and chunking. So it is clear why representations and structures are useful in the classroom. They create mental images that can be committed to long-term memory and engender deep understanding of concepts.

Misconceptions

Before we dive deeper into how to use representation and structure in lessons and then implement them into the curriculum, it might be worthwhile myth-busting some common misconceptions around representations and manipulatives:

1. Manipulatives and representations are not just for the youngest children or children who are lower attaining. Manipulatives and representations should be for all learning, so that children develop a deep and connected understanding of maths.

2. Manipulatives do not need to be used in every single lesson. Manipulatives may be useful to demonstrate a particular concept, but once this is achieved, children should build upon it and become proficient without the need for the physical equipment. This is true mastery in maths. It may well be that pedagogical choices are made as to whether children are going to use the manipulatives or whether the teacher will use them to expose a structure when modelling.

3. Giving children manipulatives does not necessarily mean that children are learning maths. Meaning does not reside in simply having tools; rather, it is constructed when those tools are used with a more knowledgeable other (the teacher).

4. Using lots of different representations is not always initially useful. In fact, this can become confusing for children. Using one representation initially may be a useful starting point (such as a part–whole model) and then, as the concept develops and that structure is secure, it may be useful to link it to another representation (such as a bar model).

Some example representations

I am sure that your curriculum already has a lot of representations included to help children understand maths, but let's look at the structures those representations actually reveal.

Number line

A number line teaches the number system (we considered this in the coherence chapter). It can help children to understand the structure of ordering numbers (including decimal fractions) and comparing numbers (including decimal fractions), both of which can include negative numbers. It can also support early understanding of addition and subtraction.

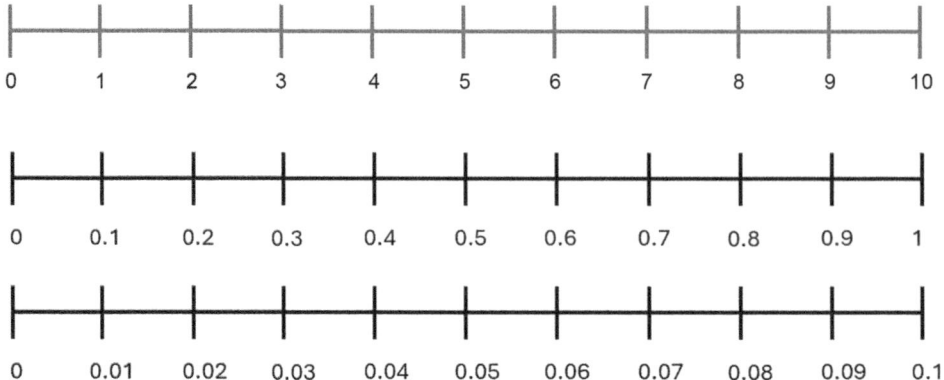

FIGURE 3.2 Stacking number lines to reveal the structure of the number system

If 6 < 9, then 0.6 < 0.9.

So if 0.06 < 0.09, therefore 0.06 < 0.05 + 0.04.

Tens frame

Tens frames help to develop number sense by providing a visual representation of numbers to ten, and they cement the base ten nature of our number system. They support subitising within this structure too. They also support counting, part–whole relationships and the composition of number (for example, they can show 6 = 5 + 1 but also 6 = 4 + 2, etc.).

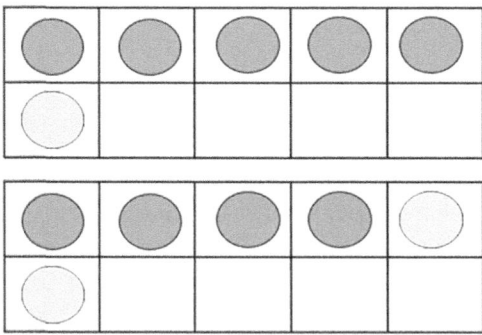

6 = 5 + 1 *and* 6 = 4 + 2 *or* 6 = 4 + 1 + 1

FIGURE 3.3 Tens frames showing how subitising and composition of number can be developed

Part–whole

A part–whole model does what it says on the tin. This representation is used right across the primary phase (including Early Years). It can be used in lots of different ways, including but not exclusive to the composition of number, parts and wholes for fractions, addition, subtraction, multiplication as repeated addition, division as sharing, decimals and so on. Again, thinking carefully about the concept that you are teaching and how a part–whole model would help to reveal the underlying maths structure of that concept is time well spent. In Figure 3.4, we show how it can be used in different curriculum areas, such as place value, addition, subtraction and in the context of money and measure.

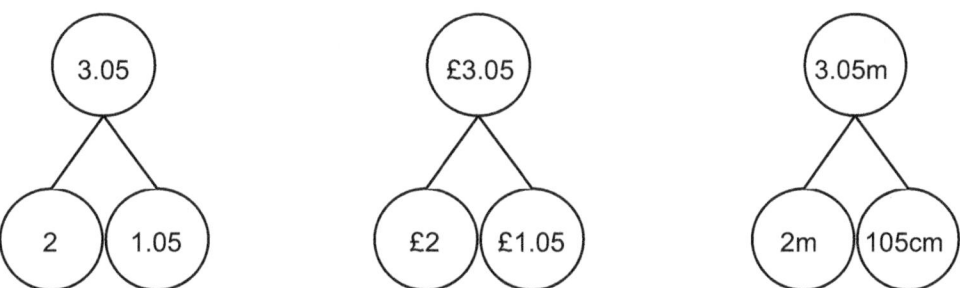

FIGURE 3.4 Part–whole models are used in every year group plus EYFS, and can link to many areas of the curriculum

Place value counters

These small discs are useful to help children to unitise (group individual items into a single unit). The discs relate to the place value and come in an array of colours. As place value runs through many areas of maths and is needed for the four operations in particular, these discs can support understanding of the value of the numbers being used, as they physically represent the numbers being used. Their power lies in the regrouping and exchanges, including knowing that ten ones are equal to one ten and that ten tens are equal to one one-hundred, and so on.

Place value counters can help to support unitising – we can see how this can link to money in Figure 3.5.

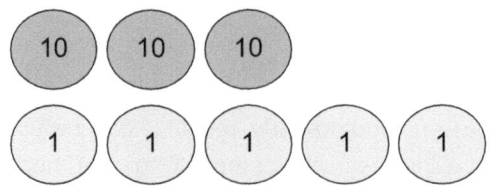

FIGURE 3.5 Place value counters can make links to money, see how 35p is represented here

Base ten

This is a structured maths manipulative designed for a specific purpose. They are used to teach the place value of the number system, in a similar way to place value counters. By physically manipulating units (ones), rods (tens), flats (hundreds) and blocks (thousands), students can visualise and comprehend the hierarchical structure of our number system, making abstract concepts more concrete and aiding in mathematical operations and understanding.

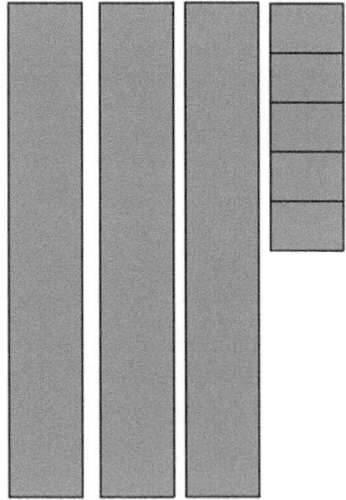

FIGURE 3.6 Base ten can support structured manipulatives and is designed to reveal the place value of numbers

Counters

These are unstructured manipulatives that can be used to reveal the structure of so many maths concepts: place value, the four operations, ratio and proportion, algebra, statistics and creating charts and so much more. They typically represent 1 but can be adapted using whiteboard pens to have any value. Careful thought should be given as to how they are used across your lesson(s) and curriculum.

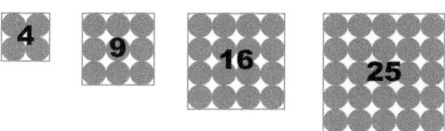

FIGURE 3.7 Using manipulatives to create square numbers

100 square

This model shows the number system in its base ten format by stacking ten rows of ten on top of each other. This visual aid is great for teaching a range of concepts, including sequencing, skip counting, place value, comparing numbers, exploring relationships, the four operations, factors, multiples and primes.

1	2	3	4	5	6	7	8	9	10
11	12	13	14	15	16	17	18	19	20
21	22	23	24	25	26	27	28	29	30
31	32	33	34	35	36	37	38	39	40
41	42	43	44	45	46	47	48	49	50
51	52	53	54	55	56	57	58	59	60
61	62	63	64	65	66	67	68	69	70
71	72	73	74	75	76	77	78	79	80
81	82	83	84	85	86	87	88	89	90
91	92	93	94	95	96	97	98	99	100

FIGURE 3.8 A 100 square

Bar model

A cousin of part–whole models, these models help to visualise problems into parts and wholes. They can be used to compare and explore relationships. Bar models promote problem-solving skills, allowing students to see relationships between quantities, ratios and the structure of mathematical operations and fractions, fostering a deeper understanding of mathematical concepts through visual representation.

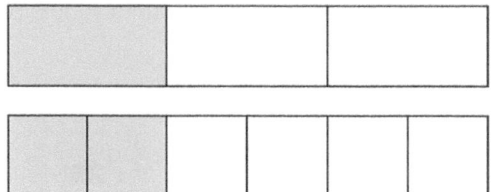

FIGURE 3.9 Bar models to reveal parts and wholes

Progression in representations

Some commercially available schemes have a wide range of representations. If we consider representations as being used to reveal mathematical structure in maths lessons, we should consider which representations are most effective in revealing the underlying structure that we are teaching. Sometimes, too many representations may be confusing for children; instead, sticking with one representation initially and deliberately, and systematically introducing other representations at points across the maths curriculum, is time well spent.

A range of schemes are used across schools. In some schools I work with, there is a high fidelity to a particular scheme. This often means that representations are progressive and build upon each other. In some schools, they adopt a 'cherry-picking' model. While this approach can work and allow for more creative freedom of teachers, it may fracture the progression of the representations in the intended curriculum. In doing so, this could impact on the coherence of the whole-school curriculum.

Table 3.1 is an example of an audit from a well-known scheme. It shows which year groups use which representations and allows all stakeholders, from maths leaders, senior leaders and teachers to governors and trust executive teams, to see, at a glance, the progression of the representations used in their curricula. It also sends a message to teachers, particularly teachers in Key Stage 2 and beyond, who may think that representations are only for younger year groups or lower attainers.

It may surprise some teachers that the use of the number line begins in Early Years and continues to be used in every single year group to Year 6 (and beyond). This is because this representation offers a valuable way in which to organise and structure the number system, from integers and negative numbers to decimals, fractions and powers of ten.

TABLE 3.1 The representations used in a well-known scheme of work

	EYFS	Year 1	Year 2	Year 3	Year 4	Year 5	Year 6
Array		X	X	X	X	X	
Tens frame	X	X	X	X	X		

TABLE 3.1 (Continued)

	EYFS	Year 1	Year 2	Year 3	Year 4	Year 5	Year 6
Dot patterns	X	X	X				
Number line	X	X	X	X	X	X	X
Part–whole	X	X	X	X	X	X	X
Rekenrek	X	X	X				
Place value counters		X	X	X	X	X	X
Place value grid			X	X	X	X	X
Fingers	X	X					
Base ten			X	X	X	X	X
Numicon	X	X	X	X			
100 square		X	X	X	X	X	
Counters	X	X	X	X	X	X	
Bar model			X	X	X	X	X
Number blocks	X	X	X	X			
Balance scales		X	X	X			
Gattegno chart						X	X

Early Years

In Early Years, we often see lots of unstructured mathematical manipulatives, such as acorns, stones and other natural resources. While these have value, we ought to consider the structures that they are actually revealing.

An important distinction for Early Years is to ensure that there is a clear understanding of structured and unstructured maths resources for children to use, with a clear intent on the concept or structure that they are revealing. Using a rekenrek in a group or whole-class session is great, but are these structured resources available in the enhanced provision areas of the classrooms? Often, structured maths resources allow children to consolidate understanding from taught sessions. In comparison, using unstructured resources such as acorns or other natural materials has value, but again, clear intent is necessary so that the focus is on maths learning.

Progression using number lines

Let's look at how this might progress from EYFS to Year 6 using the humble number line. I know that teachers have used number lines for years, but their power lies in the way in which they can be used to teach the number system. Here are some of the National Curriculum statutory requirements and ready-to-progress criteria (DfE, 2020), tracked to include a number line.

TABLE 3.2 Curriculum links and ready-to-progress criteria

	EYFS framework	Year 1	Year 2	Year 3	Year 4	Year 5	Year 6
National Curriculum statutory requirements (DfE, 2013)	Have a deep understanding of numbers to 10, including the composition of each number.	Identify and represent numbers using objects and pictorial representations, including the number line, and use the language 'equal to', 'more than', 'less than' (fewer), 'most' and 'least'.	Identify, represent and estimate numbers using different representations, including the number line.	Identify, represent and estimate numbers using different representations.	Round any number to the nearest 10, 100 or 1,000.	Round any number up to 1,000,000 to the nearest 10, 100, 1,000, 10,000 and 100,000.	Round any number to the required degree of accuracy.
Ready-to-progress criteria (DfE, 2020)	Reason about the location of numbers to 10 within the linear number system.	Reason about the location of numbers to 20 within the linear number system, including comparing with <, > and =.	Reason about the location of any two-digit number in the linear number system, including identifying the previous and next multiple of 100 and 10.	Reason about the location of any three-digit number in the linear number system, including identifying the previous and next multiple of 100 and 10.	Reason about the location of any four-digit number in the linear number system, including identifying the previous and next multiple of 1,000, and 100, and rounding to the nearest of each.	Reason about the location of any number with up to two decimal places in the linear number system, including identifying the previous and next multiple of 1 and 0.1 and rounding to the nearest of each.	Reason about the location of any number up to 10 million, including decimal fractions, in the linear number system, and round numbers, as appropriate, including in context.

Let's consider how this might look for each year group.

EYFS

Reason about the location of numbers to 10 within the linear number system

When children start Reception, we need to be selective about the representations to which we are exposing them. Which ones are going to serve them well into their whole school career? The number line is a representation that reveals the structure of the number system. Using it allows children to order numbers and count them forwards and backwards. This is the starting point that can unlock the entire structure of our number system.

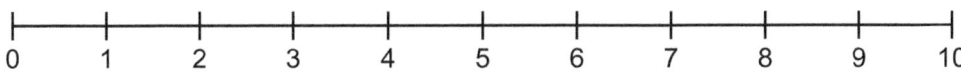

FIGURE 3.10 A number line showing 0–10

Year 1

Reason about the location of numbers to 20 within the linear number system, including comparing with <, > and =

Using number lines from 0–10 and then 0–20 would be useful to build upon the understanding of 10 from EYFS. Counting forwards and backwards can be useful, as can describing two numbers based on which is largest and smallest. It can also be useful to identify the midpoint of 5 on a 0–10 number line and the midpoint of 10 on a 0–20 number line. 'Stacking' number lines can be useful here too: showing children a number line with clearly marked intervals from 0–10 and then, underneath it, a blank number line from 0–10 can support children in making connections between both of these representations. Similarly, having a 0–10 number line and, underneath, a 10–20 number line allows children to see the pattern of numbers between 10 and 20 and the pattern of the ones digits.

Year 2

Reason about the location of any two-digit number in the linear number system, including identifying the previous and next multiple of 100 and 10

It can be useful to introduce a number line using a bead string with 100 beads on it. Children will be able to see the beads (intervals) and reason about which number they are showing. Look for the midpoint on a number line from 0–100 and then look for midpoints between multiples of 10 and ask children what they notice. Once children are gaining fluency, we might ask them to identify the previous and next multiple of 10. Children should also be given practice with a number line with intervals and a blank number line, to reason about the location of numbers. This can be effective if they are 'stacked' so that children can make the connections. Again, referring back to the bead string can be effective in supporting understanding here.

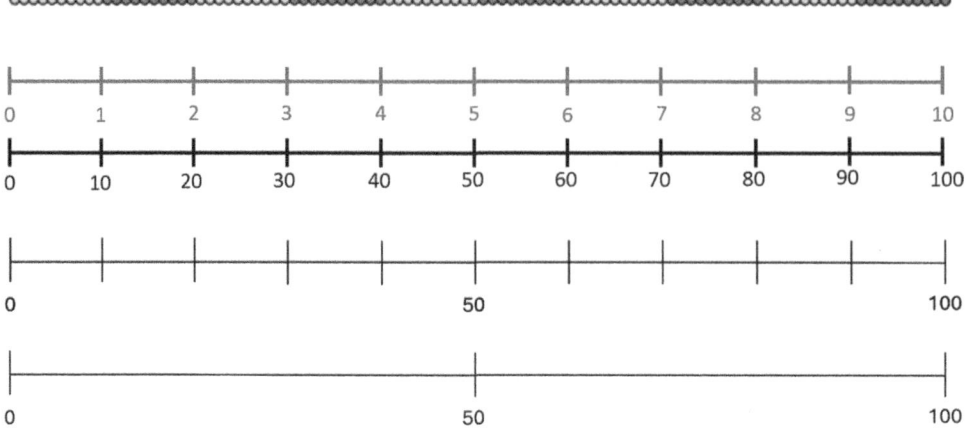

FIGURE 3.11 Making connections between a bead string and number lines – notice how the number lines progress to being almost blank so that children can reason about the position of numbers in the linear number system

We can also begin to use the same representation but start to think about it differently. For example, we may present the number line as vertical. Here, we can ask children to identify which numbers are being pointed to and allow them to reason, using number lines with intervals and blank number lines.

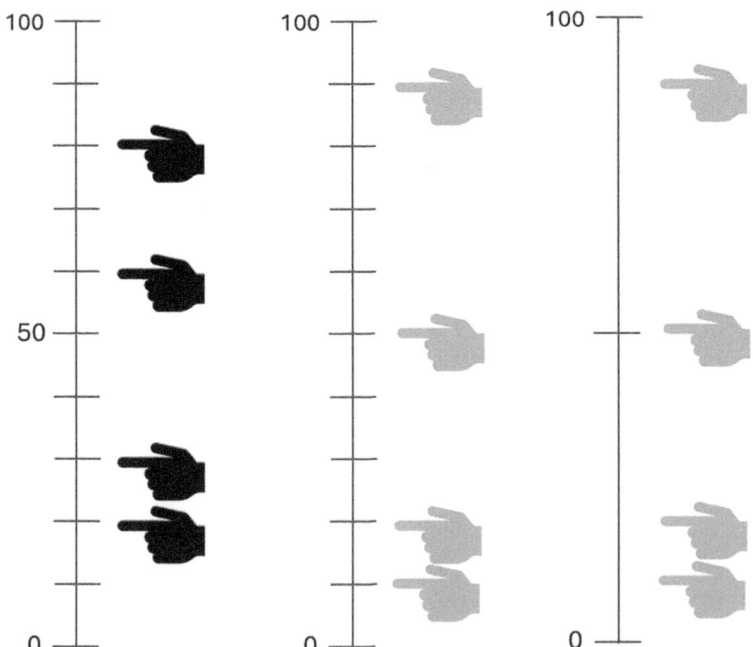

FIGURE 3.12 Number lines presented vertically, gradually removing the intervals until they are almost blank; midpoints can be shown as a point of reference

Once children have learned the structure of the number system, we might then add contexts such as the same example shown in Figure 3.13.

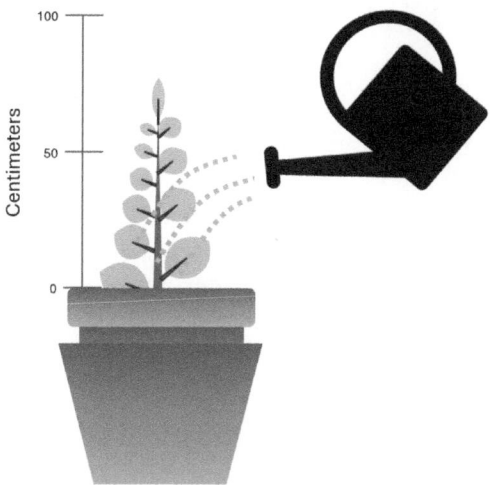

FIGURE 3.13 A vertical number line in the context of measure

We could push this representation a little further again. Sticking with the idea of a number line and the position of numbers on a number line, we might introduce a curved number line too. This can allow us to start to teach in the context of mass, for example.

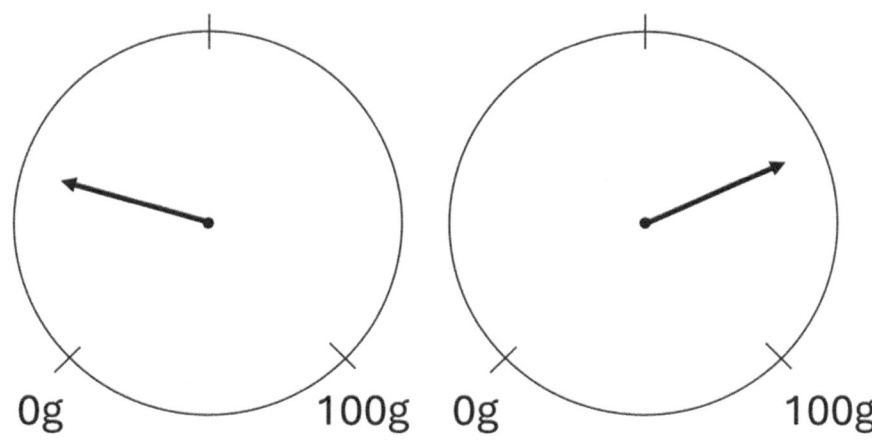

FIGURE 3.14 A number line (curved) in the context of measure

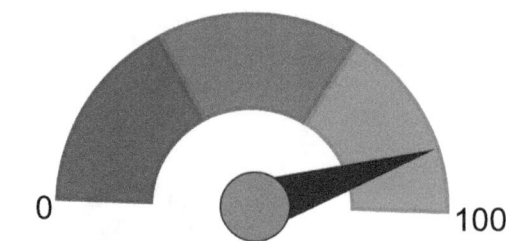

FIGURE 3.15 A number line presented in the different context of a gauge – this could link to weight, speed or capacity

In the chapter on coherence, we discussed coherence in representations. All of these representations reveal the structure of the number system using a number line. Hopefully, you can see how these can link to other areas of the curriculum, including their usefulness with fractions of amounts, decimals and percentages later down the track.

Year 3

Reason about the location of any three-digit number in the linear number system, including identifying the previous and next multiple of 100 and 10

We can now map this core representation looking at three-digit numbers. We might start by reminding children about 0–10 number lines and the connections that we can make in 0–100 number lines. We can then extend this thinking using 0–1,000 number lines. Identifying the midpoint is crucial in order to anchor children's reasoning. We can then consider given numbers and estimate where they could be placed, based on the previous and next multiple of 100. We might then make connections to other representations of the same number lines but using different contexts. This could be using a measuring jug with a vertical number line on it, weighing scales with a circular measuring representation or revisiting temperature. All of these representations require an understanding of the structure of the number system. The representations that we use are the vehicle to revealing these structures. In all of these examples, identifying the midpoint is a good starting point to anchor thinking, as are all of the previously mentioned activities but simply done in a slightly different way.

Year 4

Reason about the location of any four-digit number in the linear number system, including identifying the previous and next multiple of 1,000 and 100, and rounding to the nearest of each

This continues in much the same way as in Year 3 but uses a 0–10,000 number line. We have already introduced previous and next multiples of a number to a given context, and so we can really begin to focus on this and introduce the concept of rounding, using the number line representation to structure children's thinking. I hope that you can see how this has built up from a 0–10 number line.

Year 5

Reason about the location of any number with up to two decimal places in the linear number system, including identifying the previous and next multiple of 1 and 0.1 and rounding to the nearest of each

By now, children have been systematically exposed to the number line representation and should be developing a strong sense of the number system. By Year 5, we may track back to the 0–10 number line and explore decimals. Having blank number lines is useful too, and asking where the same number might be placed on each number line can help with contextual reasoning. Again, rounding to the nearest whole number introduces the concept of rounding.

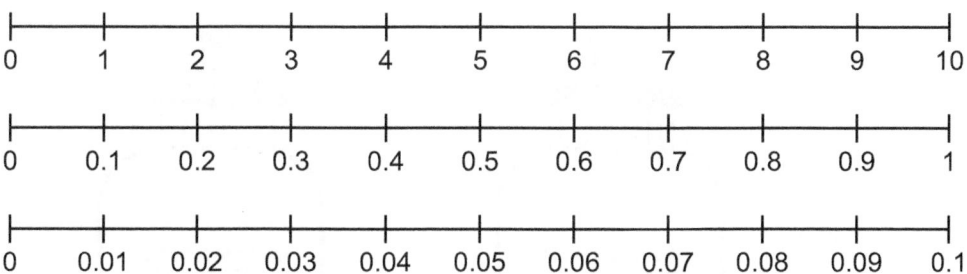

FIGURE 3.16 Stacked number lines reveal the connections (powers of 10) in the number system

Year 6

Reason about the location of any number up to 10 million, including decimal fractions, in the linear number system, and round numbers as appropriate, including in context

You can now see how this concept builds over time. It avoids relying on unconnected and abstract rhymes, such as: '1, 2, 3, 4, go to the multiple before, 5, 6, 7, 8, 9, go to the next one on the number line.' This gimmick misses out so many other opportunities for children to really reason deeply and carefully. In Year 6, children can use their vast knowledge of the number system to round to any given degree of accuracy, including on vertical and horizontal number lines and in the context of measuring jugs or scales, see examples in Figure 3.17 and Figure 3.18 below.

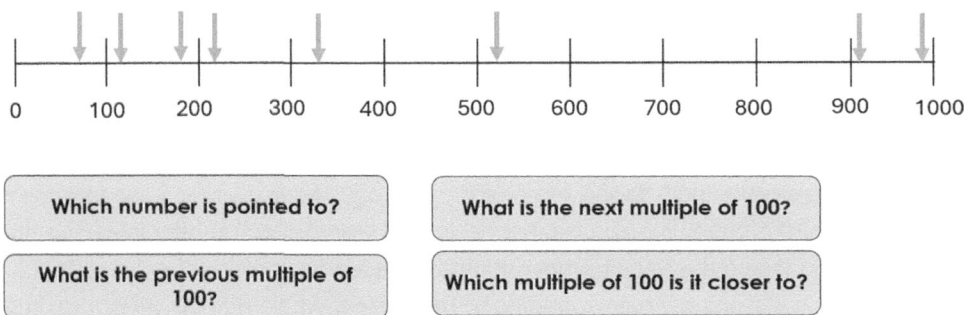

FIGURE 3.17 Arrows pointing to a number to allow children to reason about rounding

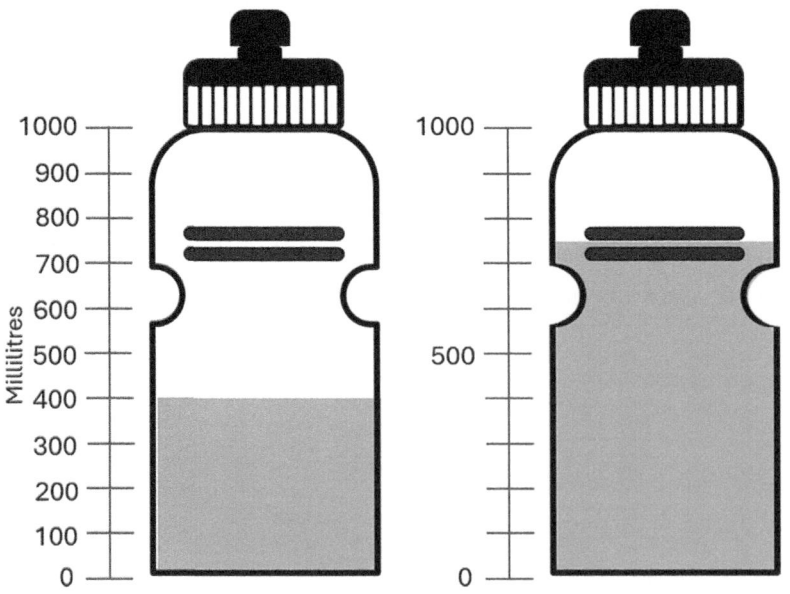

FIGURE 3.18 Rounding to the nearest millilitre in context using vertical number lines

Clearly, representations and structures cannot stand alone but work with the other elements of teaching for mastery. For example, here we have described a *coherent* journey using the representation of a number line. Cherry-picking resources may not lead to this level of coherence in a curriculum, and so it ought to be carefully, systematically and deliberately planned out.

We spoke about S-planning in Chapter 2. This can be a valuable staff meeting, inset day or team meeting to have with colleagues. Drawing out the structure of your curriculum on an S-plan and tracking when a specific representation is used and for what concept can be time well spent. This can work in two ways: it can develop teacher pedagogy and understanding of what, how and why they are teaching a concept and it can bring greater coherence to the curriculum if colleagues are equipped with knowing why particular representations are most suited to particular concepts.

You may want to take some time now to pause and reflect on where this concept may begin in EYFS and Year 1 and how it develops across the school.

Part–whole models

Let's look at a different concept. I'll start with the Year 6 objective below, and I wonder whether you may pause for a moment and consider where this concept might begin in EYFS and how it develops through school to Year 6 (see Table 3.3). Another way of saying this might be: if this is the expectation for the end of Key Stage 2, how do we get children there? This is a great activity to begin to consider a particular representation and what it looks like across school.

Objective: Read, write, order and compare numbers up to 10,000,000 and determine the value of each digit

TABLE 3.3 Progression in part–whole model representation

EYFS	Year 1	Year 2	Year 3	Year 4	Year 5	Year 6	Key Stage 3
Partitioning numbers (composition to 5 and 10)	Place value of numbers to 20	Place value of two-digit numbers	Place value of three-digit numbers	Place value of four-digit numbers	Place value of numbers to two decimal places	Place value of any digit in numbers to 10,000,000	Structure of the number system

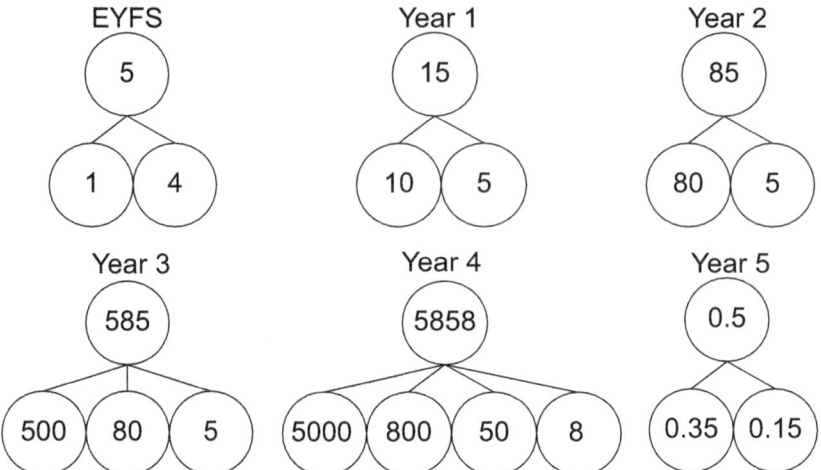

FIGURE 3.19 Part–whole models can be used flexibly to deepen the structure of the number system

Figures 3.19 and 3.20 are just examples; there are many ways in which we can develop and deepen the part–whole model so that children can partition numbers more flexibly and reveal the structure of numbers.

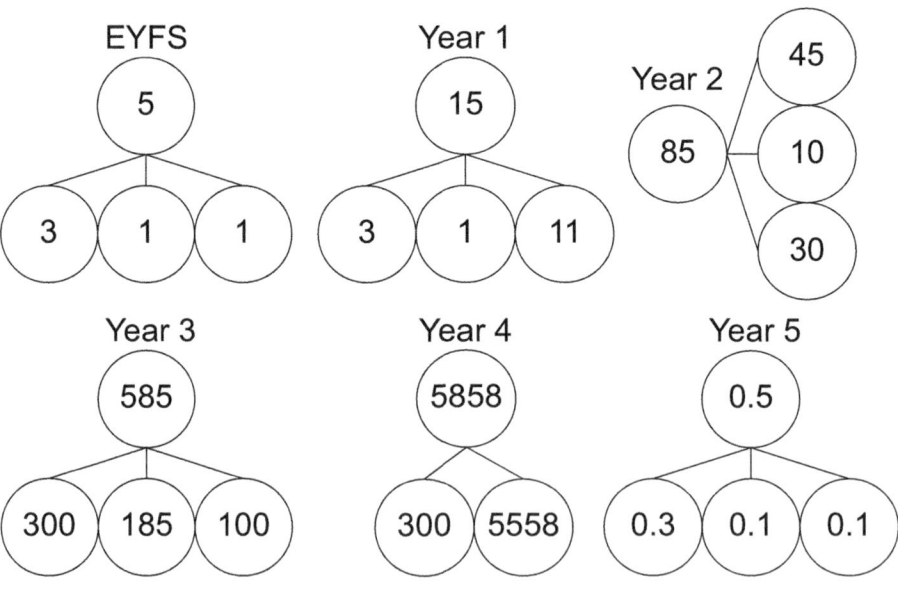

FIGURE 3.20 Partitioning numbers in different ways to develop strong number sense and number flexibility

These examples demonstrate the structure of composition of numbers using place value. We may now see the power of this representation in revealing the structure, but perhaps focusing on addition and subtraction (not to mention the focus on specific vocabulary for each of the parts of the part–whole model) or even the composition of fractions. When using core representations across the curriculum, knowing where they begin and the structure that they reveal is essential to teaching concepts effectively. Alongside this, we might introduce key vocabulary for reasoning. In this example, we show that if we swap the values of the subtrahend and difference, the minuend stays the same.

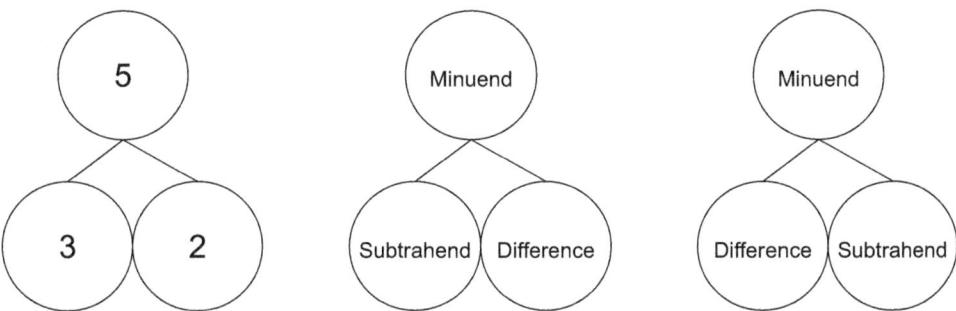

FIGURE 3.21 The vocabulary linked to calculation can be used to reveal their structure

I've sat through many lessons that have attempted to teach children the formal algorithm for addition and subtraction. Most lessons start with well-intentioned thinking around the place value of each column, and some of the better lessons use place value counters to show what this number looks like using a place value grid. Often, as the lesson progresses, children become stuck because their knowledge that ten 1s makes one 10 is not secure, and lessons can quickly descend into a messy web of counters being flicked around, place value grids having smiley faces drawn on them and the teacher resorting to an abstract and didactic battle cry of 'Here's how to do it, just remember!'. This often leads to misguided teaching that relies on gimmicks, such as placing a 1 or a milk bottle on the doorstep, etc. If the Year 3 objective was prioritised and taught before, during and after the introduction of calculating, the cognitive load would be reduced when thinking about exchanging ten 1s for one 10.

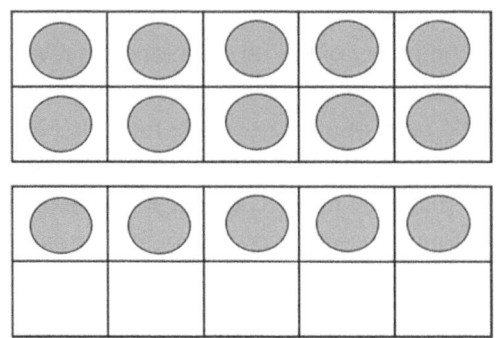

FIGURE 3.22 A tens frame revealing the composition of 15

In Figure 3.22 we can see that 15 is ten 1s and five 1s combined or one 10 and five 1s. This flexibility with number is crucial in order for children to understand key concepts in the curriculum (see Table 3.4).

TABLE 3.4 Progression in from EYFS to Year 6

	EYFS	Year 1	Year 2	Year 3	Year 4	Year 5	Year 6
Ready-to-progress criteria (DfE, 2020)	Begin to develop a sense of the number system by verbally counting forwards to and beyond 20, pausing at each multiple of 10	Count forwards and backwards within 100		Know that ten 10s are equivalent to one 100, and that 100 is ten times the size of 10; apply this to identify and work out how many 10s there are in other three-digit numbers	Know that ten 100s are equivalent to one 1,000, and that 1,000 is 10 times the size of 100; apply this to identify and work out how many 100s there are in other four-digit multiples of 100	Know that 10 tenths are equivalent to one 1, and that 1 is 10 times the size of 0.1 Know that 100 hundredths are equivalent to one 1, and that 1 is 100 times the size of 0.01 Know that ten hundredths are equivalent to one tenth, and that 0.1 is 10 times the size of 0.01	Understand the relationship between powers of 10 from 1 hundredth to 10 million, and use this to make a given number 10, 100, 1,000, 1 tenth, 1 hundredth or 1 thousandth times the size (multiply and divide by 10, 100 and 1,000)

Some schemes have too many representations all at once. This is not a problem once children are familiar with a representation, but I think that, in most cases, we should teach one thing and teach it well. Dotting between base ten, place value counters, abstract representations and pictorial representations can be somewhat confusing for children and may be counterintuitive to the depth of their understanding. We should think carefully about which structure a representation or manipulatives would show.

Have you used a tens frame before? If so, I'd like you reflect on the ways in which you have used it (this makes a great staff meeting on representations, with tens frames and double-sided counters ready for staff to use and demonstrate). I'm sure that many Early Years and Key Stage 1 teachers are thinking of the multiple ways in which they have used tens frames. I would take a punt that Key Stage 2 teachers are less familiar with tens frames – after all, they can only take us so far, right? Wrong! They're very versatile and can be used in every single year group, from Reception to Year 6 (and beyond).

We learn using a base ten system, perhaps because we have ten digits connected to our hands (fingers and thumbs), and so lots of our mathematical structures are connected to the base ten system. Representations can reveal these structures.

Tens frames are one of the representations that begin in Reception and are used in every single year group in primary. Knowing their use and what structures they expose is important. In Figure 3.23, we can see how the idea begins in Reception with numbers to 5. Children may do a lot of work filling in half a tens frame and learning the composition of number to 5 (5 + 0, 4 + 1, 3 + 2, 2 + 3, 1 + 4, 0 + 5). This could be matched with fingers (finger gnosis). We can see how this idea builds into Year 1 with numbers to 10 and a full tens frame being one 10. Children are learning that ten 1s make one 10 – huge concepts that will pay off in dividends when they need to exchange with column addition and subtraction in Key Stage 2. By Year 2, children are using the tens frame to think about strategies for subtraction: 13 − 5 = 13 − 3 − 2. The importance of bridging through 10 becomes very clear.

By Year 3, children are learning that ten 1s equal one 10 and ten 10s equal one 100. Into Year 4, children learn that ten 100s equal one 1,000. Year 5 takes this further, still using a tens frame including decimals, learning that ten tenths equal one whole and ten hundredths equal one tenth. By Year 6, children are exploring the structure of the tens frame using all powers of 10, from hundredths to 10 million. Using a tens frame is powerful, and seeing its use in every year group shows how teachers can have an awareness of what has gone before, what should happen in a particular year group and where it progresses to at a later stage. This is inextricably linked with subject knowledge and how to teach concepts and ideas well.

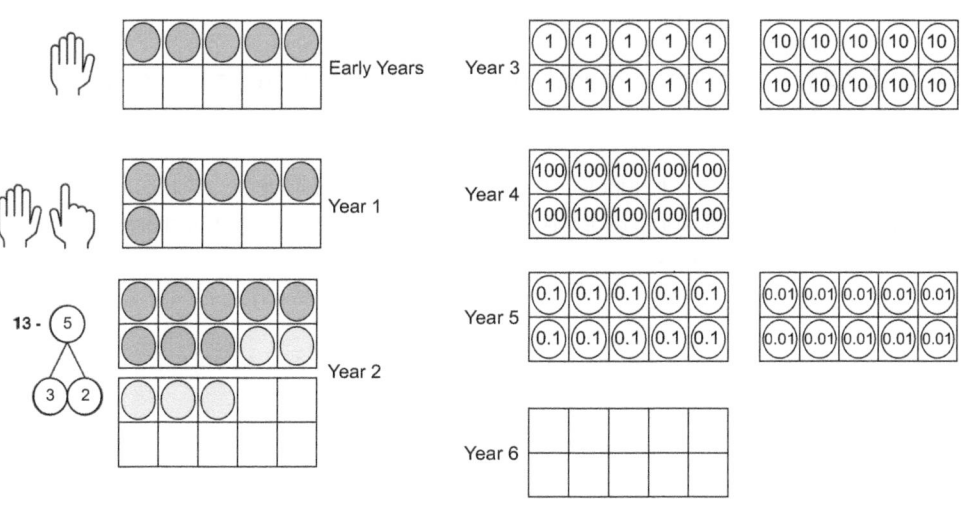

FIGURE 3.23 Progression in representations (tens frames)

Additive structures

Representations matter when showing structure. The part–whole model in Figure 3.24 shows the additive structure of addition. Manipulating this structure by adding addends in a different order and revealing the commutative structure of addition is an important learning point, as this can be generalised to all addition. The equation of addend + addend = sum shows the relationship between three numbers. In primary, many of us will be familiar with writing fact families, such as $13 + 8 = 21$. Exploring this to reveal the structure is important.

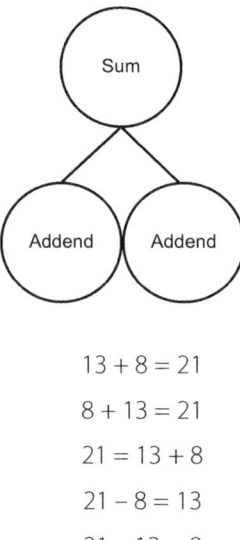

$$13 + 8 = 21$$
$$8 + 13 = 21$$
$$21 = 13 + 8$$
$$21 - 8 = 13$$
$$21 - 13 = 8$$
$$13 = 21 - 8$$
$$8 = 21 - 13$$

FIGURE 3.24 A part–whole model showing the additive structure, using example of $13 + 8 = 21$

We can see how this leads nicely on to the structure of inverse and writing related equations in Figure 3.25, and that if we swap the values of the subtrahend and difference, the minuend stays the same.

Sum	
Addend	Addend

Minuend	
Subtrahend	Difference

Minuend	
Difference	Subtrahend

FIGURE 3.25 Bar models showing the structure of inverse

So, we can now think about what each part of the equation of a subtraction means so that we can explore the structure of subtraction. In Figure 3.26, we can see the structure of subtraction in a different way. Representations, again, reveal the structure of mathematics.

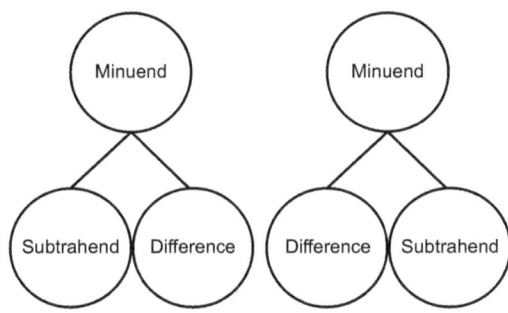

FIGURE 3.26 Part–whole models showing the structure of inverse

Multiplicative and division structures

Multiplication has its own structure too. We want children to master this structure beyond knowing that factors are commutative. We can consider quotative (grouping) and partitive (sharing) as the main structures.

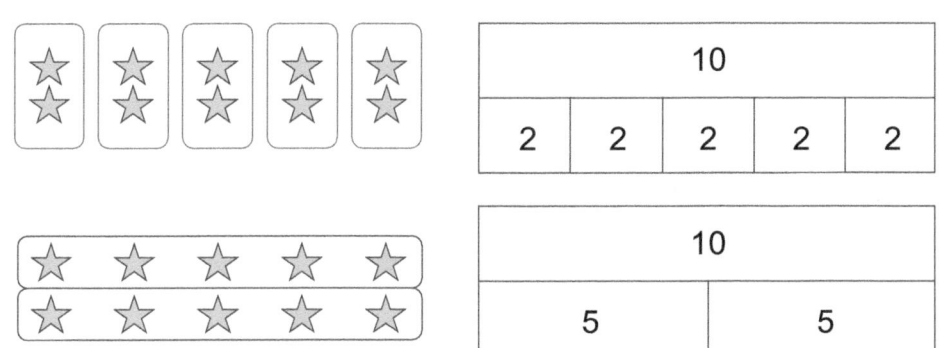

FIGURE 3.27 Representations showing quotative and partitive structures

Can you see which is which using the representation in Figure 3.27? The first example shows quotative division, as ten stars have been divided into five equal groups, whereas the second example shows partitive division, as we can see that the size of each part is five stars. The structures of the expressions here are different: the quotative structure is $10 \div 5 = 2$ and the partitive structure is $10 \div 2 = 5$. These structures help us to understand division in different contexts and provide insights into how we can approach problem-solving. Quotative division focuses on the number of groups, whereas partitive division focuses on the size of each share. It can help to ask children, alongside these representations, what each number represents. Try a stem sentence, such as: '10 is the dividend and it represents 10 stars; 5 represents the number of equal groups (the divisor); 2 represents the number of stars in each group (the quotient).' Again, we can see how the representations support the thinking around this important mathematical structure.

Fraction structures

Understanding fractions and what they represent is important in maths. We must start with fractions as a part of a whole. This is key to all understanding of fractions. We can represent this in infinite different ways.

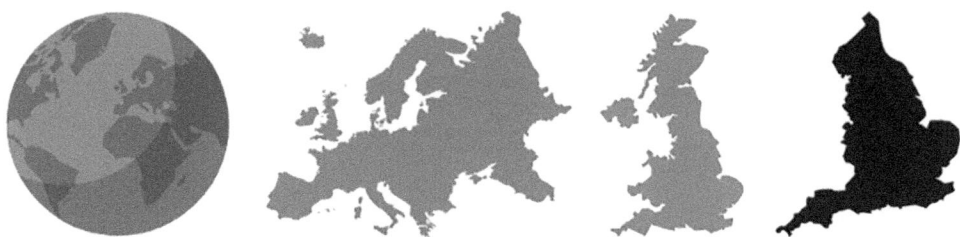

FIGURE 3.28 Fractions shown in a non-maths context of parts and wholes

Here we can start teaching the relative size of things too, such as the world being a whole and the oceans being part of that whole, or Europe, the United Kingdom or England being the whole.

FIGURE 3.29 Fractions shown in a non-maths context of parts and wholes

Again, each of the items in Figure 3.29 can be described as the whole and something that is a part of it can be described as being part of the whole. We also need to teach the concept of equal and unequal parts fairly early on, as this is key to understanding the structure of fractions. Once secure, we might then start to show children fraction notation ($\frac{1}{2}$ again), drawing attention to the structure of the whole having two equal parts, with one of those parts selected. This takes time and plenty of practice. We must be mindful of the type of representations that we show children, so as not to only offer a limited exposure to a concept's structure. The representations in Figure 3.30 show the idea that equal parts may not look the same. Here, we show thirds in different ways to build up the concept.

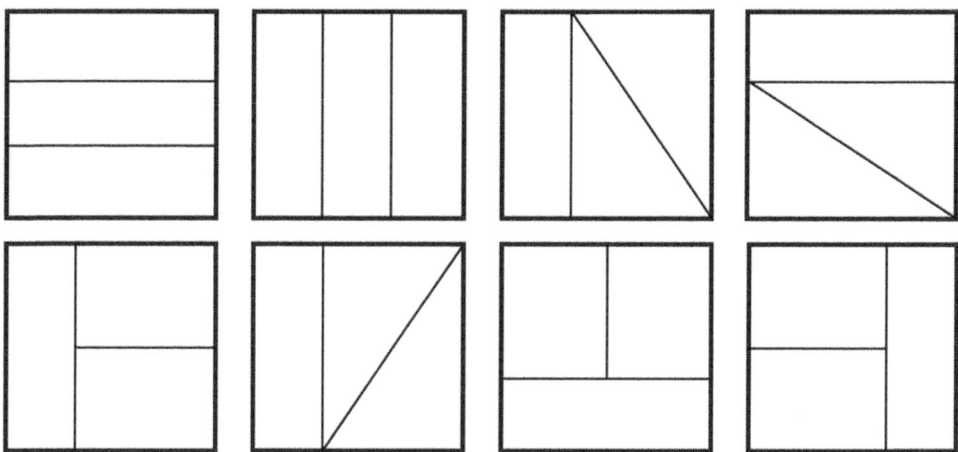

FIGURE 3.30 Representing thirds in different ways to build the idea of what thirds can look like

We can then begin to compare unit fractions by looking at the same whole divided into different parts, but we should also consider and actively teach some common misconceptions. The representations of $\frac{1}{4}$ and $\frac{1}{2}$ can be confusing, due to the relative sizes of the whole. It's an important teaching point that the wholes must be the same size in order to compare fractions.

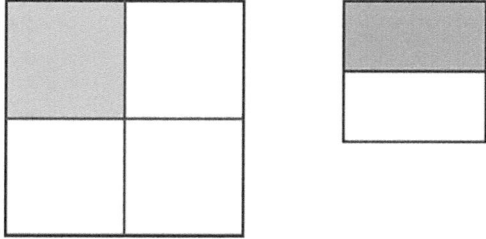

FIGURE 3.31 Representations drawing out the structure of fractions

The representations in Figure 3.31 are revealing the structure of the fraction concepts that we are teaching here in a powerful way.

Unpicking the structure of fractions is important so that children know how maths works. Fractions are closely linked with division. In fact, the obelus (the name of the division symbol) is a fraction bar with two unknown amounts shown on it as dots: ÷. This is because of the relationship between fractions and division. This becomes particularly important when looking for equivalent fractions. Look at the examples below. What do you notice? And, importantly, what might children do to work this out or show this?

$$\frac{2}{4} \quad \frac{17}{34} \quad \frac{3}{6} \quad \frac{5}{10} \quad \frac{250}{500}$$

You might be able to generalise that they're all equivalent to $\frac{1}{2}$. But how do you know that? What if we represented the relationship like this instead: $\frac{a}{b}$? Can you now explain the relationship between a and b – between the numerator and the denominator? Where a is half of b, we can be sure that the fraction is equivalent to a half. Now we can generalise this for fractions equivalent to $\frac{1}{3}$ or $\frac{1}{4}$ or any unit fraction. Looking at the examples below, which are equivalent to $\frac{1}{4}$?

$$\frac{2}{8} \quad \frac{3}{12} \quad \frac{5}{25} \quad \frac{20}{100} \quad \frac{17}{68} \quad \frac{25}{100}$$

Did the structure of equivalents and the relationship between numerators and denominators help with your reasoning?

Of course, representations can show us why this structure works: $\frac{1}{2}=\frac{2}{4}=\frac{3}{6}=\frac{4}{8}$ (notice the deliberate change in representation).

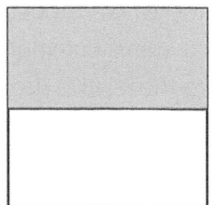 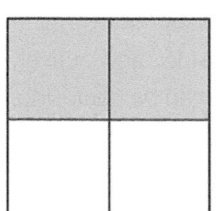 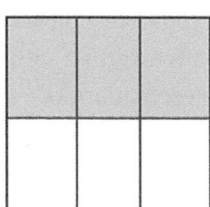 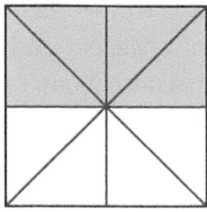

FIGURE 3.32 Representations drawing out the structure of equivalent fractions

Figure 3.33 shows the same equivalence in a different way to Figure 3.32.

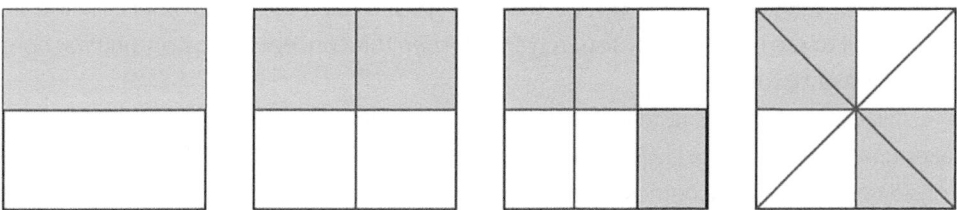

FIGURE 3.33 Representations drawing out the structure of equivalent fractions

Embedding representations in classrooms is key to teaching for mastery. It moves beyond passive acceptance of mathematical rules and conventions to deepening understanding of how and why concepts work. We must, of course, remember that $\frac{1}{2}$, $\frac{2}{4}$, $\frac{3}{6}$ and $\frac{4}{8}$ are also representations. They are abstract and should be used alongside pictorial and concrete representations to unlock a world of understanding.

Representations can be used to stimulate thinking. Simply showing a representation and asking what children can see, what they notice and what they wonder can be an effective way in which to get lessons started. While I'm on the subject of starting lessons, I want to address learning objectives – more specifically, sharing them at the start of each maths lesson. As mentioned previously, a few years ago, there was a popular TV show called *Take Me Out* in which women stood on an illuminated podium awaiting a suitor, who would arrive in the love lift. Based on initial impressions, the women would switch off their lights to show their disinterest in the unfortunate soul who was parading around. Learning objectives can work a bit like this. As soon as we state that 'today we will be learning about equivalent fractions', we can sometimes see the lights behind children's eyes going out. I'm not suggesting that we shouldn't have a clear idea of what children should learn – to the contrary, we should be very clear about what we want children to learn. But how we do that and design lessons can go a long way towards changing mindsets. Perhaps, instead of sharing the learning objective, we might show two representations of equivalent fractions and ask children to discuss what they notice. We can then begin to explore the idea of equivalent fractions for the duration of the lesson, using a range of representations that show equivalence (and do not show equivalence) as the lesson progresses.

I'm a big fan of giving children activities in books with limited (if any) text. Sticking with the earlier example, we might give children representations of $\frac{1}{2}$ and $\frac{2}{4}$ and ask them to draw a few more fractions equivalent to $\frac{1}{2}$. We might then develop tasks to include non-examples (more of this in Chapter 6 on variation) so that children build strong ideas of what equivalence is and is not. We might vary the shape of the whole (circles or triangles) to build conceptual understanding too. Once children have gained conceptual fluency, they have developed mental images of this concept and can move on to other thinking. It's not difficult to see how we might stretch and deepen this concept using non-unit fractions and equivalents, such as $\frac{2}{3}, \frac{4}{6}, \frac{6}{9}$.

Using representations to reveal mathematic structures, like anything worth doing well, takes time. The deliberate choices made around what to use in lessons should be carefully considered. Published schemes do some of the heavy lifting about this, but opportunities can be missed if teachers do not know the structures that the representations can reveal. In visualising the structures through careful and deliberate use of representations, children can visualise to internalise their thinking of key mathematical concepts.

Table 3.5 is a synopsis of a lesson that I taught in Year 5. The focus was on the area of compound shapes.

TABLE 3.5 Year 5 lesson synopsis on the area of compound shapes

Small step	Representation	Analysis
I started by showing children two shapes and asking what was the same and what was different. I told children that one side of the square was 8 cm. I asked them to then tell me what they could find out.		This allowed children to elicit the fact that each side was equal. After a few moments, I was able to discuss perimeter and area to guide children's thinking on what they had recently been learning about. This focused their thinking to calculate the perimeter and area of each shape and describe what was the same and different.
I asked children to draw lines in their book accurately.		This allowed me to assess who could draw accurately with a ruler. As the session would require this skill to reveal the concept of area of compound shapes, I could target my support to children who needed help with drawing accurately.

TABLE 3.5 (Continued)

Small step	Representation	Analysis
On coloured card, I asked children to draw the following shapes and then cut them out carefully. These were then stuck into books as a point of reference for children.	5 cm, 5 cm; 3 cm, 3 cm; 5 cm, 7 cm; 3 cm, 4 cm	Children had four different shapes. I asked them to calculate the area of each one, reminding them of how they had done this in previous lessons. I then asked them to put two of the shapes together and tell me the combined area of the shapes. I told children that this was a compound shape and they had calculated the area of compound shapes.
The following representations were shown on the screen. I asked children to calculate the areas A and B and then combine. I showed the children one example by drawing a scale model on card and 'sticking it together' to show why we combine both areas of a compound shape.	A: 4 cm, B: 2 cm, 6 cm, 3 cm; A: 10 cm, 6 cm, B: 8 cm, 7 cm	The removal of the card for children allowed them to make connections to the pictorial representation. I used my version of the compound shape on card to model how to calculate missing sides. I could also fold the compound shape that I had made to show strategies of how to split the shape. This was displayed at the front of the class as a point of reference.
These representations were stuck into books and children calculated the area of compound shapes.	3 cm, 5 cm; B: 2 cm; 4 cm, A	The representations were kept the same but shown in different orientations to ensure high levels of success when practising the skill.

(Continued)

TABLE 3.5 (Continued)

Small step	Representation	Analysis
	3cm 8cm · B 4cm · A 7cm	
The lesson continued using a 'ping pong' approach, where I would teach an example and children would practise. Towards the end of the lesson, I stuck in the representation that I started with and gave children one length of the square. I asked children to write down the areas of the compound shapes and explain what was the same and what was different about the representations.		

Stem sentences as a structure

I'm often asked whether I have a definitive bank of stem sentence that link to the curriculum. The short answer is no. This completely misses the point of a stem sentence. Stem sentences can be implemented in two distinctive ways in your classrooms and schools:

1.   To practise key vocabulary: '3 is an **addend**, 4 is an **addend**, the **sum** is 7.'

2.   To generalise: 'A fraction is a part of a whole.'

Stem sentences can be used to support children's reasoning and thinking in lessons, but they also support mathematical structure in the case of vocabulary practice. The stem sentence 'addend + addend = sum' can be used to teach the commutative nature of addition. Figure 3.35 shows the addends 3 and 4. The first image shows 3 + 4 = 7 and the second image shows the addends swapping, so 4 + 3 = 7.

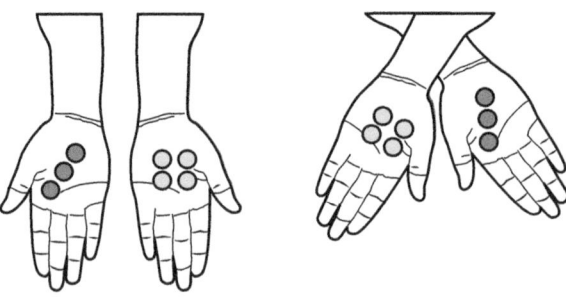

FIGURE 3.35 Representations showing the commutativity of addition – when the addends are swapped, the sum stays the same

This is a generalisation of 'addend + addend = sum' or 'sum = addend + addend'.

This can then lead to a clear explanation of the commutativity of addition, so: $3 + 4 = 4 + 3$.

Here, we have the interplay between the use of the representation to show the structure and the stem sentence to reinforce the structure that we are teaching, so that it becomes clear to children. The understanding in the example above can then be generalised to any numbers and strengthened by using three or more addends to equal a sum. Furthermore, it will stay with children well into Key Stage 2 and beyond, when they are adding $0.3 + 0.4 = 0.7$. By Year 6, children are leaning into what has been started in Year 1 when deriving related facts. For example, $891 = 243 + 648$ and so $89.1 = 24.3 + 64.8$ and so $8.91 = 2.43 + 6.48$ and therefore $6.48 + 2.43 = 8.91$ (commutative) and so $6.49 + 2.43 = 8.92$ (adding one hundredth to an addend impacts the sum by adding one hundredth).

With the prevalence of 'off the peg' schemes and the conflicting priorities of teachers, there is a danger that we take representations for granted without really considering the structure that they are revealing. The number line is a powerful reminder of how we systematically and deliberately teach the number system, including fractions and decimals. Similarly, part–whole models are used from Early Years and well into Key Stage 3. Knowing how and why these are used for concepts is important so that you can draw attention to the structure (partitioning structure, for example) when exploring the composition of number. Again, using S-plans from EYFS to Year 6 (and beyond) and adding objectives to them for a particular stand of maths can be a great way to start thinking about how objectives progress. Revisiting these and adding on core representations and how they progress from EYFS to Year 6 can be effective professional development, so that staff can appreciate where their year group fit into the bigger picture. As we will see, when we teach with structure at the front of our thinking, we really get to the heart of teaching mathematics well.

Key takeaways

The thoughtful integration of representations and mathematical structures is foundational to effective mathematics instruction. This chapter has provided an argument against traditional methods that rely on superficial tricks and rote memorisation, which often fail to cultivate a deep understanding of mathematical concepts. Instead, it champions the use of well-chosen representations and manipulatives to uncover and reinforce the underlying structures of mathematics.

The distinction made between representations and manipulatives is particularly noteworthy. Representations, as visual or symbolic depictions, and manipulatives, as physical tools, both serve to make abstract concepts tangible and comprehensible. This approach demystifies mathematics, making it accessible and engaging for students across all levels of proficiency. Importantly, the myth that manipulatives are only for young or struggling learners is laid bare, advocating instead for their use as essential learning aids throughout all stages of education, regardless of ability.

Using representations can avoid common misconceptions. For instance, rounding is not taught as a random (yet catchy) rhyme in lessons but is instead taught conceptually to demonstrate how the concept works. I have also argued that mere possession of these tools does not guarantee learning; meaningful interaction guided by knowledgeable instruction is crucial. This perspective encourages a balanced and thoughtful use of educational resources, ensuring that students can internalise and build upon their learning experiences.

I felt that it was important that a significant portion of this chapter was dedicated to illustrating how various representations, such as number lines and part–whole models, can be utilised effectively. I hope that the practical examples demonstrate how these tools can guide students from basic numerical understanding to more complex concepts, like decimals and fractions. This progression underscores the importance of coherence and continuity in curriculum design, where each new concept builds logically on prior knowledge. We can see clearly that coherence in the progression of representations is also an important consideration when teaching for mastery.

It is essential for educators to have a robust understanding of the mathematical structures that they aim to teach because this knowledge enables them to select the most appropriate representations and manipulatives, facilitating a deeper learning experience for their students. The role of professional development and collaborative planning and lesson design should be underscored and systematically planned out across a school year (or school years for long-term sustained change). In doing so, we can promote a culture of continuous improvement and shared expertise among teachers.

The approach and practical examples here will go a long way towards equipping children with the skills and knowledge that they need to succeed.

4 Mathematical thinking

We are always thinking about one thing or another, but how we think frames our understanding and our ways of viewing the world around us.

'Watch your thoughts for they become your words, watch your words for they become your actions, watch your actions for they become habits, watch your habits for they become your character, watch your character for it becomes your destiny. What we think, we become.'
(Outlaw, 1977)

What does this mean for maths? How we describe maths and the words that we use when teaching it shape the way in which children view maths. Some well-known published schemes may be unintentionally unhelpful. Many of them separate 'fluency', 'reasoning' and 'problem-solving' into separate parts, which can lead to what is effectively a three-part lesson under a different name. This may send the wrong message, as these three things should work together and with each other. In order to reason and solve problems, children need to be fluent, and in order to be fluent, children need to be able to apply fluency when reasoning and solving problems.

The National Curriculum states that children should:

'reason mathematically by following a line of enquiry, conjecturing relationships and generalisations, and developing an argument, justification or proof using mathematical language' (DfE, 2021).

I think that maths can be one of the most creative and imaginative subjects. We want children to hypothesise, conjecture, wonder, confirm, deny, imagine, critique and everything in between to reveal the beauty of the maths world. This chapter will outline how we may do that in practical terms.

When we ask genuine questions, we can ignite something that requires us to think carefully and deeply. Take the example in Figure 4.1 overleaf. It is best to look up these flags online so you can see them in full colour. I want you to ponder and think carefully about what the answer could be. I'm going to hold back on the answer for some time, as giving you the answer limits your ability to think imaginatively and creatively.

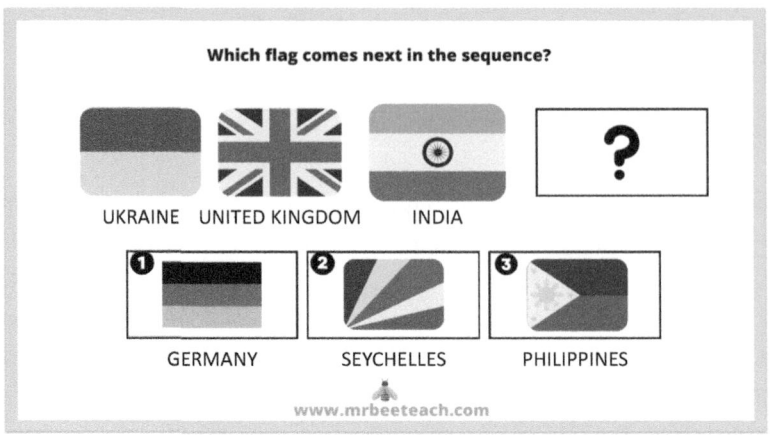

FIGURE 4.1 An example of what reasoning can look like and *feel* like

Your brain is hard-wired to look for patterns, links and connections. Doing this activity may have caused some cognitive conflict in your brain, and the subsequent result is your brain trying to find calm among the cognitive chaos. I've shared this task in rooms full of both children and adults, and it is always thrilling to hear people's reasoning and rationale behind their thinking. Now that you've had time to consider this question, if I told you the answer is 2, would that support your thinking? It may well allow you to sharpen your reasoning around why this is the case, but by not revealing the 'how' or 'why', you will continue to keep thinking. In this example, this type of question allows all children (and adults) to access the task. I am sure that many of you noticed that each subsequent flag increases by one colour as part of your reasoning, and so the answer must be flag 2. How do we capture this type of enthusiasm in every single lesson that we do?

Many schemes have high-quality questions and tasks for children to complete. Being selective and deliberate about the tasks that we give children is an important consideration. Looking behind what the answer is to *why* the answer is what it is allows children time to imagine, prove, disprove, conjecture, reason, make connections and think mathematically.

Let's imagine that a scheme of work has Figure 4.2 as a suggested task for children to do as part of a worksheet. The fractions have been very carefully chosen (more of this in Chapter 6) and they're a great idea for some independent practice.

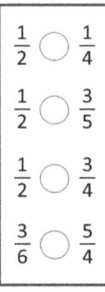

FIGURE 4.2 An example of questions where we want children to notice something

What might we want children to do in their books? How can we rethink this activity so that children are showing their understanding and thinking deeply about the structure of the questions?

Figure 4.3 shows the same activity completed in two different ways. Children learn what they attend to. The design of what we want children to do is an important part of how we might get children to think in classrooms. I go through the process of what children will be doing when planning lessons. I consider which representations I might want children to draw in books, which key vocabulary might be useful for children to use in their reasoning and how this could be completed in books. We might even supercharge these activities with the scaffold of key vocabulary printed next to the tasks for children to use in their written reasoning: equivalent and mixed numbers would be useful here for children. You can see that there are still further opportunities here to show $\frac{3}{6}$ as equivalent to $\frac{1}{2}$ or $\frac{5}{4}$ as $1\frac{1}{4}$, etc.

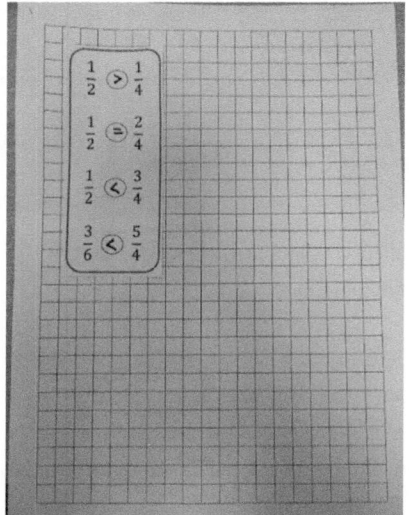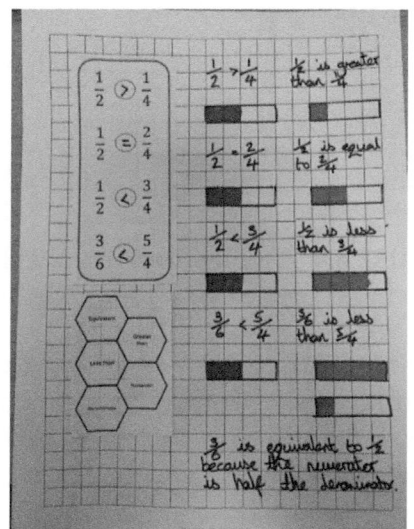

FIGURE 4.3 A maths book with completed tasks showing the depth to which children can go with well-designed tasks

Notice how this example could also lead us to the generalisation that when comparing unit fractions, the larger the digit in the denominator is, the smaller the fraction. It really does depend on what we want to draw attention to.

Thinking carefully about what we want children to attend to and to learn is important in mathematics. We can apply the same idea across much of the maths curriculum, with carefully designed tasks that promote rich learning for all children. For example, when teaching times tables, many of us will chant with children so that they remember the facts and can recall them quickly to apply to problems. But let's look at how we might get children to make mathematical generalisations with times tables and the types of activities that promote it. Figure 4.4 shows two activities with very different outcomes.

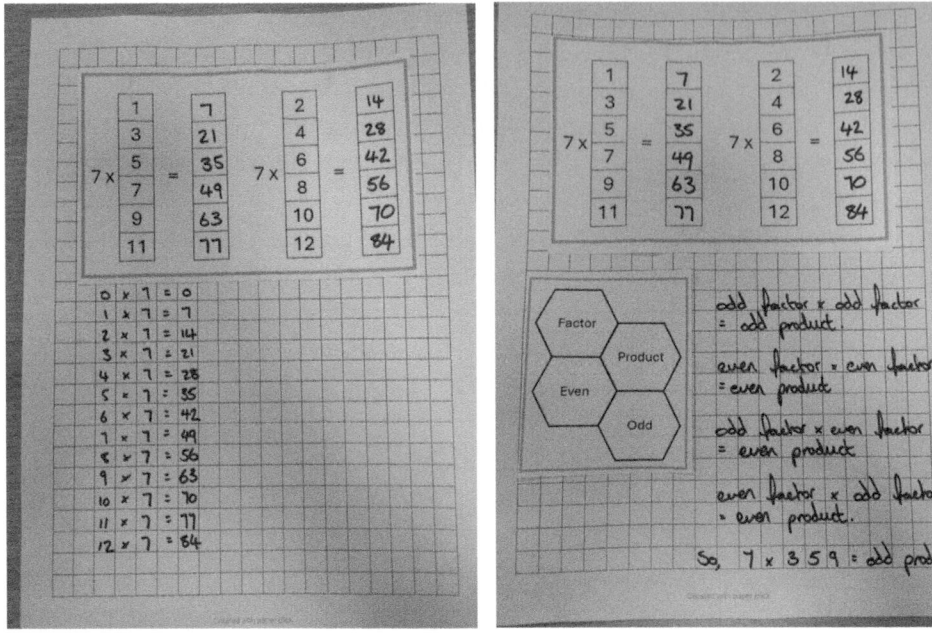

FIGURE 4.4 Another maths book with completed tasks showing the depth to which children can go with well-designed tasks

In the first example, there are many missed opportunities, and here there is limited value apart from memorising the facts (which is important), but the second example goes deeper. The activity splits numbers into odd and even and then requires children to make generalisations about the types of products when different types of factors are multiplied. These generalisations are universal rules to which children can come back to again and again, because if they know that odd factor x odd factor = odd product (such as 7 x 9 = 63), then we can deduce that 7 x 359 should result in an odd product, as should 7 x 593. The second activity can then be adapted for children who finish quickly for them to sort given calculations into odd or even products, so that they suit the rule. There really are limitless ways in which this could be developed in the moment within a lesson.

Reasoning without limits

We're really drilling down to what reasoning is: following genuine lines of enquiry, making conjectures about relationships, generalising and developing argument, justifying, hypothesising and proving (and disproving). But what does it *feel* like to reason? Well, take a look at the image in Figure 4.5. What do you notice? What do you wonder? What can you see?

FIGURE 4.5 An example of an image to prompt maths reasoning, see www. mrbeeteach.com for colour version

Activities like this *feel* genuine and don't have a closed answer. They're accessible for all children. I have used this image from the youngest of children to adults and I'm always startled by the lines of enquiry, conjectures, relationships, generalisations, arguments, justifying, hypothesising, proving (and disproving) that I witness. I use this a lot in my Year 6 class, as it covers multiples, factors, primes, square numbers, etc., all so beautifully. It is, of course, important not just to give unrelated tasks to children in lessons but to ensure that children have the components to reason and solve the problems that you give them – for instance, in this example, multiples, factors, primes, square numbers and recall of times tables all need to be secure to access it deeply. Much of the lesson is about discussion and talk: reasoning and thinking creatively and flexibly. When there is a natural lull in the class, I might then interject with some structured questions around multiples or prime numbers to guide thinking. Only when this discussion has been completed, I might ask children to wonder what a number (I usually go for 48) might look like. I get children to draw it in their books and reason why, using vocabulary of multiple, factor, prime, square, cube and so on.

The tasks that we give children to do are important. Their design and potential to support mathematical thinking are critical. This is often an area that can become a bit of an afterthought for teachers, with their focus being on what they will be doing with the whole class and regurgitating something from a scheme or a task that somebody else has created. There is nothing inherently wrong with using tasks from published schemes; in fact, some of them are really excellent. Their downfall is often in the way in which children respond to them in writing. A sentence or one word just doesn't cut it for reasoning. We have to let genuine questions breathe and flow. We have to discuss and wonder. To do this, we must anticipate the type of response that we want children to produce. I'm a huge fan of a visualiser in the classroom (more on this later) to explicitly teach children how to draw mathematical models and annotate them to support their reasoning. In doing so, we're unpicking the structure of questions and getting to the heart of the matter (and often the underlying maths structures).

In designing tasks like this, we also avoid the terrors of ability groupings. Instead, we personalise and deepen children's thoughts through our structured questioning.

- *What do you notice?*
- *Tell me about multiples of 2.*
- *What would number X look like? Why?*
- *Show me what number 101 would look like.*
- *How would number 50 and 500 be the same? How would they be different?*
- *Show me how number 24 and number 48 would be the same. Show me how they would be different.*

Now, there are other things that we must teach in the curriculum. But I'm trying to put a sharp point on task design and how deep we can take learning in lessons. It's far better to do few things well than lots of things poorly. I'm often alarmed in school classrooms at how quickly children get through questions in books. Extension activity after extension activity seem to be stuck in books in a 'race to the top'. This approach transmits lots of messages to children: any maths question should be able to be completed quickly, or if you can't answer it quickly, you are not good at maths. This is a particular type of 'school maths' and is far removed from what mathematicians spend their time doing. Many of them devote much of their working lives to solving problems and certainly do not rush to get finished.

Embedding this across school takes time and effort. Teachers need to see it in action. Inviting teachers, in small groups, to observe you can be a powerful and transformative model of professional development; it can lead to critical action and positive changes in practice.

We want children to remember what we teach them. We must remember that children are novice learners, and while what we teach them may build on previously taught concepts, much of the learning is new to them. I want you to think of the different types of numbers that we teach children (integers, negative, fraction, decimals, prime, composite and so on). Now we might consider the types of numbers that are not taught in schools. There is a whole world of numbers obscured from our curriculum that evade our thoughts well into adulthood and perhaps for all of our lives. All of these types of numbers have cousins and distant relations that make up our number system – we just do not know about them because they are not taught, so we do not think about them: vampire numbers, narcissistic numbers, cake numbers, happy numbers, evil numbers, pronic numbers and repunit numbers. Let's take a closer look at one of these: vampire numbers.

Vampire numbers

Vampire numbers are a special type of number that can be expressed as the **product** of two other numbers, where the digits of those **two numbers are rearranged and put together** to form the original number.

To better understand **vampire numbers**, let's look at an example:

1,260 = 21 x 60

In this example, we see that the digits of 21 and 60 can be both rearranged and multiplied together to give us 1,260. This makes 1,260 a vampire number. The term 'vampire number' is used because they can be seen as numbers that 'suck the life' out of the two numbers that make them up.

But not all numbers can be vampire numbers. For a number to be considered a vampire number, it **must have an even number of digits** and the two numbers that make it up must have the **same number of digits**.

Let's look at a few more examples of vampire numbers:

1. 1,827 = 21 x 87

2. 102,510 = 201 x 510

3. 104,260 = 260 x 401

4. 125,460 = 204 x 615

As you can see from the examples above, vampire numbers can be quite large and have a lot of **factors**. In fact, some vampire numbers can have **multiple pairs of factors** that can be rearranged to form the original number.

One interesting thing about vampire numbers is that they are quite rare. In fact, there are only a few hundred known vampire numbers in existence. This makes them a fascinating topic for mathematicians and number enthusiasts alike. So, you see, the potential of thinking in creative ways can unlock a world of understanding, and maths can feel imaginative and creative and genuine. I hope that the reasoning and potential of vampire numbers is clear in the framework of teaching for mastery (despite their absence from the curriculum). I used to share these numbers with children after SATs and as transition activities to get children thinking creatively and imaginatively.

Task design

Children learn what they attend to.

I started teaching in the last days of the National Numeracy Strategy. This document gave specific timings and structure to lessons, with some suggested representations and use of interactive teaching resources. However, there was limited support in terms of what children would attend to during their tasks. After teaching thousands of maths lessons in my career and subsequently sitting in hundreds of maths lessons, the tasks that we give children are really important to get right. Designing logical and coherent tasks that allow children to succeed is essential to effective teaching and learning and for children to make progress with mathematical concepts and ideas. I am sure that many of you are using schemes of learning that provide some support in the form of worksheets. Worksheets can be a starting point but they can cognitively overload children, and maths can quickly become a race through a series of loosely related tasks. However, I think there is merit in using some of these (or some of your own) to develop rich tasks that allow children to show their depth of understanding. Imagine that you have been teaching fractions and children have had practice with comparing fractions. In this lesson, you have taught children to compare unit fractions with different denominators. You're now ready to set children off with an activity to do in their books. Let's take the activity in Figure 4.6 as an example.

Only a fraction of each rod is shown. Which rod is longer?

FIGURE 4.6 An example of a question on fractions that requires deep thinking

It may be easy for children to look at this and complete it after a few moments by answering correctly $\frac{1}{9}$, but what if we reimagined this task and gave it a bit more depth? If we go beyond the answer and ask for reasoning to prove or show their thinking, it now becomes an activity in revealing the structure under the card.

$\frac{1}{9}$	$\frac{1}{9}$	$\frac{1}{9}$	$\frac{1}{9}$	$\frac{1}{9}$	$\frac{1}{9}$	$\frac{1}{9}$	$\frac{1}{9}$	$\frac{1}{9}$

$\frac{1}{8}$	$\frac{1}{8}$	$\frac{1}{8}$	$\frac{1}{8}$	$\frac{1}{8}$	$\frac{1}{8}$	$\frac{1}{8}$	$\frac{1}{8}$

FIGURE 4.7 Showing the maths and representing the maths can be a powerful way to get children to think

By encouraging children to show which of these rods is longest, the children get to practise drawing (and therefore thinking about) what this structure looks like. It also provides valuable assessment for the teacher as to whether children can recognise equal sizes of each part and the number of equal parts. We may then add further depth to this representation by asking children to produce some written reasoning alongside the representation, to explain their thinking – perhaps something like: '$\frac{1}{8} > \frac{1}{9}$; however, as the parts are equal sized, the $\frac{1}{9}$ rod is longest.'

We can now deepen this a little further, so children can then draw and represent their original statement that $\frac{1}{8} > \frac{1}{9}$. This requires children to show that the total lengths of the rods are equal, as the whole has been divided into nine equal parts and the same whole has then been divided into eight equal parts, so those parts have a different size relative to each other.

$\frac{1}{9}$	$\frac{1}{9}$	$\frac{1}{9}$	$\frac{1}{9}$	$\frac{1}{9}$	$\frac{1}{9}$	$\frac{1}{9}$	$\frac{1}{9}$	$\frac{1}{9}$

$\frac{1}{8}$	$\frac{1}{8}$	$\frac{1}{8}$	$\frac{1}{8}$	$\frac{1}{8}$	$\frac{1}{8}$	$\frac{1}{8}$	$\frac{1}{8}$

FIGURE 4.8 Making the wholes equivalent changes the thinking around this problem

Again, we could ask children to use the inequality symbols alongside this, to show that $\frac{1}{8} > \frac{1}{9}$.

So, where do we go next? The opportunities are limitless. We could heavily scaffold this so that children repeat some of the activity but the depth of the activity changes somewhat, as in Figure 4.9.

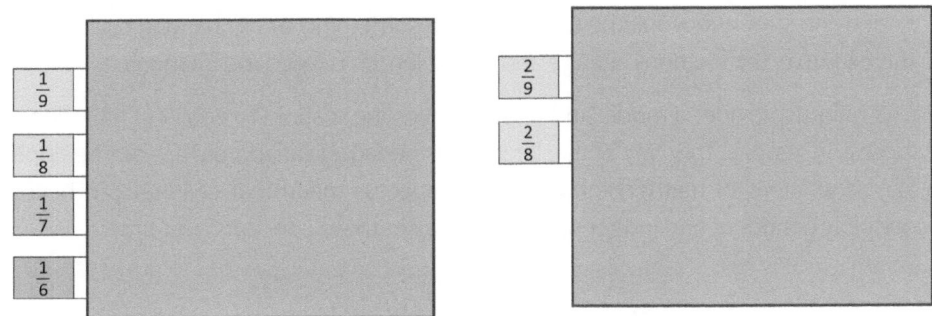

FIGURE 4.9 Same structure, different depth of thinking in these examples

Here we have thought carefully about the tasks and how we want children to explain the structure of the maths that they are learning.

'I do, we do, you do' approach

In a wider sense, using an 'I do, we do, you do' approach allows children to clearly see the expectations of how to present their learning. This approach can also allow you to clearly model your thinking out loud and so communicate the structure of these problems. Imagine that the three slides in Figure 4.10 are part of your lesson presentation slides. Each activity/representation is shown on individual slides (or animated to appear after each click).

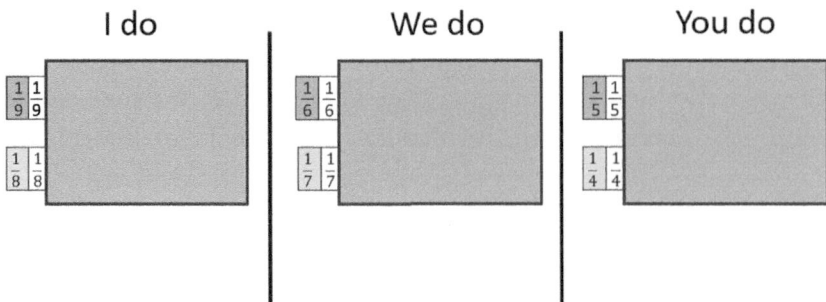

I do	We do	You do
$\frac{1}{9}$ $\frac{1}{9}$	$\frac{1}{6}$ $\frac{1}{6}$	$\frac{1}{5}$ $\frac{1}{5}$
$\frac{1}{8}$ $\frac{1}{8}$	$\frac{1}{7}$ $\frac{1}{7}$	$\frac{1}{4}$ $\frac{1}{4}$

FIGURE 4.10 An 'I do, we do, you do' approach can support thinking through worked examples

- **I do:** You model this to the children and they watch as you represent and show children your thinking in how to work this out. This can be done by thinking out loud and showing children how they could represent this in their books.

- **We do:** All children (perhaps this is stuck in their books) follow your direct instruction step by step, to show children how to be successful. This may be useful under a visualiser displayed on the board (with your very own maths book to work in too).

- **You do:** This would be stuck in children's books for them to try independently. The underlying structure would be the same, with something small changed each time. In this example, the fractions change and so the length of each rod changes.

This technique provides a model and then removes the scaffold to engineer high degrees of success in lessons. This way of teaching – this pedagogical approach – can be applied to almost all areas of maths. High degrees of success mean that children can become increasingly confident and independent, and master these concepts over time.

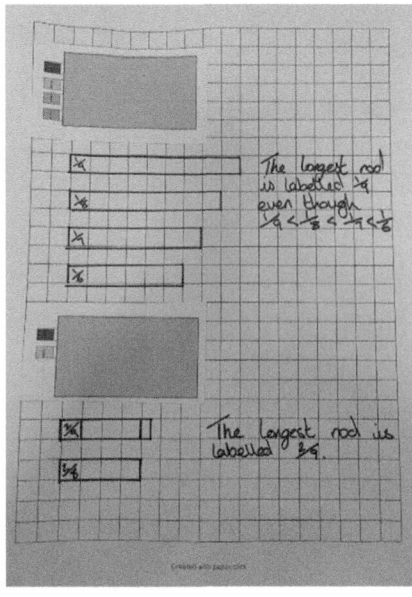

FIGURE 4.11 Examples of what success with reasoning in books could look like

Connecting thinking

Let us now turn to another idea so that you may appreciate how this can work and be implemented across school. Here we consider an example from Key Stage 1. We know that number lines are used across school from Early Years to Year 6 (and beyond). They're a core representation that help us to teach about the structure of the number system. Ordering numbers and placing them on number lines depend on the context of the numbers. For example, when children learn to label number lines and the position of numbers, we might begin by showing them something familiar, such as a 0–10 number line.

0   5   10

FIGURE 4.12 A 0–10 number line with the midpoint identified as 5

Identifying the midpoint is important here, as we can then begin to deduce where other numbers would come. This becomes particularly important when we introduce blank number lines later on. We might then tap into children's knowledge about 1 more than 0 and 1 less than 10, or children may count in 1s to complete the number lines.

We can use this as a basis to the lesson and begin to develop tasks very carefully. We could guide children's thinking about the start numbers, end numbers and midpoints, asking what they notice. We might then begin to teach where particular numbers would be placed using the intervals to help us.

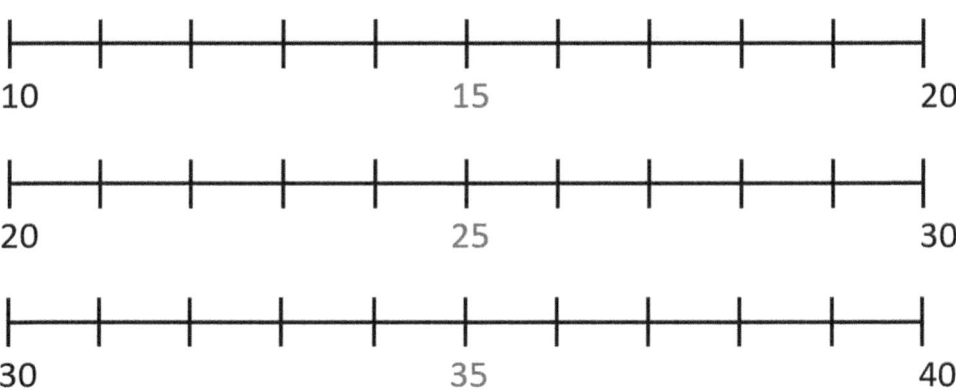

10   15   20

20   25   30

30   35   40

FIGURE 4.13 Stacked numbers with deliberate choices made about the start, end and midpoints to draw attention to the structure of the number system

We might then design similar questions and ask where specific numbers would go on blank number lines. For example, where would '47' go on each of the number lines in Figure 4.14?

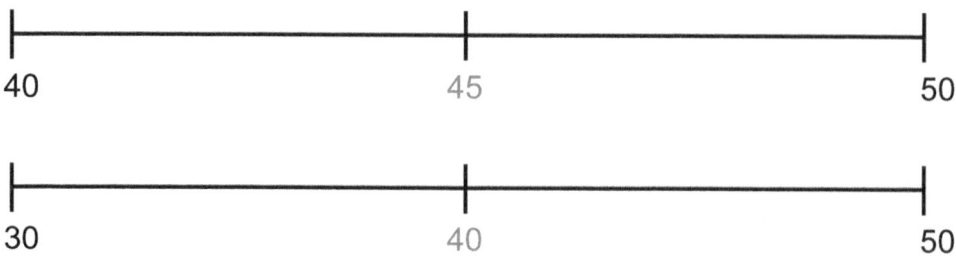

FIGURE 4.14 Stacked numbers with deliberate choices made about the start, end and midpoints to draw attention to the structure of the number system

Tasks designed in this way support children's understanding of the number system and make later concepts, such as rounding, conceptually more accessible for many learners. Looking for the links, connections and how the number system works builds and creates strong visual models of the structure of our number system. The image in Figure 4.15 shows how we might manipulate the number line representations to draw out the structure of the number system.

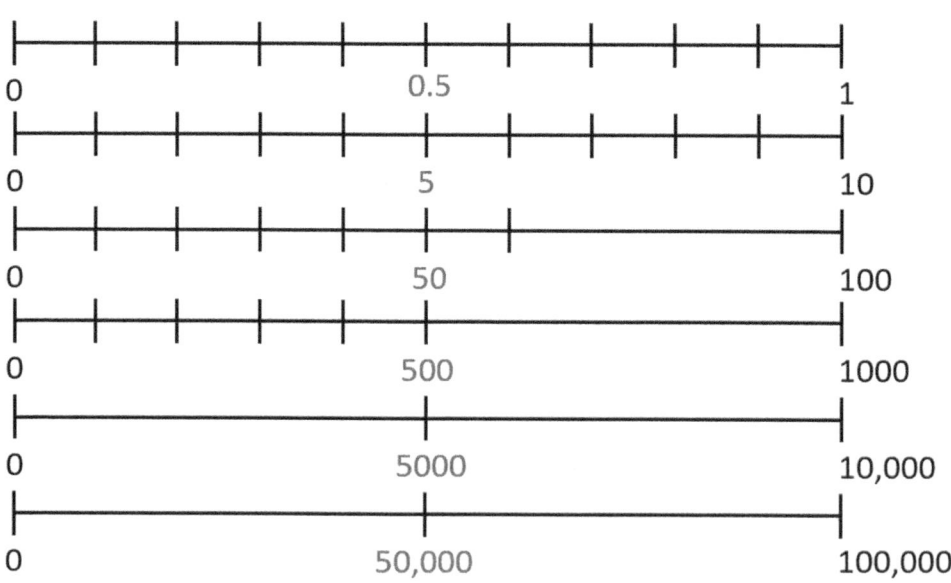

FIGURE 4.15 Stacked numbers showing the structure of the number system

This highly structured approach can go a long way to supporting progress while reducing maths anxiety. The implications for rounding, decimal number systems, fractions and concepts that children encounter later in their life are clear… and it all starts with the strong foundations of the humble 0–10 number line.

Problem-solving

Solving problems is central to mathematics. As we have seen, when the problems seem genuine, it can lead to children asking, questioning, justifying, wondering, hypothesising and changing their minds. One of the issues that we can have is that we often ask children to attend problems that are contrived or convoluted (Key Stage 2 SATs, anyone?). There may be no intrinsic motivation to solve page after page of problems wrapped up in words or strange contexts. The context or situation of problems can mask the underlying mathematical structure, and so children need to recognise this underlying structure, often in a particular order, to solve problems.

Let's consider the example in Figure 4.16 through the lens of the underlying mathematical structure:

Here are three identical rectangles. Calculate the area of the shaded triangle.

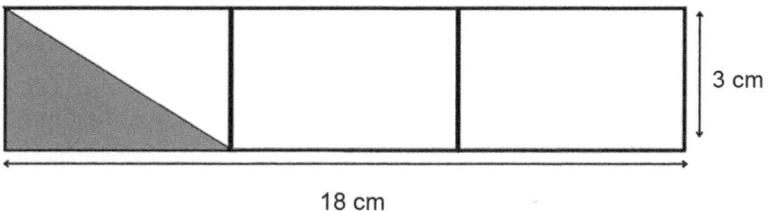

FIGURE 4.16 An example of a problem that children could be asked to solve. What are the component parts needed to solve this?

This problem requires children to make connections to previously taught concepts and combine them. Children need to identify the component parts of the problem and then combine them in order to answer the question effectively. It needs breaking down into small, manageable steps, focused on the components of the problem. Here, children would need to recognise that they should divide 18 cm by three rectangle lengths, so that the length of one side may be found. Children would then need to multiply 6 cm and 3 cm together to calculate the area of one of the rectangles. This requires them to recall length x width = area of rectilinear shapes. Children should finish by halving the area of one rectangle to calculate the area of the shaded triangle (9 cm²) – remembering the squared symbol too! Alternatively, children could show 18 cm x 3 cm and then divide this by six triangles.

Once the components have been taught, this worked example is left displayed on the screen or a small whiteboard and children may have a go at a routine problem with the same surface structure.

Here are six identical rectangles. Calculate the area of the shaded triangles.

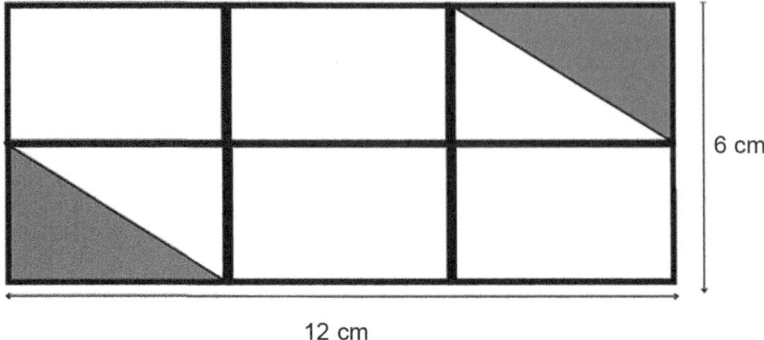

6 cm

12 cm

FIGURE 4.17 Same structure, different depth when problem-solving

At this point, I might print out a scale model of this for children to cut and 'see' that both triangles make $\frac{1}{6}$ of the whole.

We could give children some similar examples to have a go at so that they can solve routine problems. Perhaps the triangle would be in a different place or there would be different measures used. As we have taught children the structure of this problem, after some routine practice, we might show children something that is non-routine and see whether they can make the cognitive leap to solve questions such as the problems in Figures 4.18 and 4.19.

Here are six identical rectangles. Calculate the area of the shaded part.

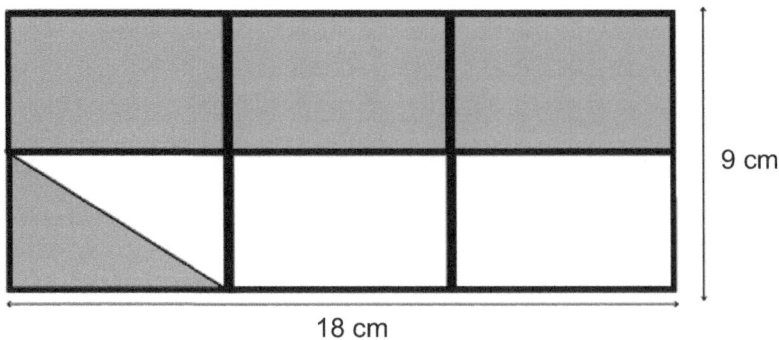

9 cm

18 cm

FIGURE 4.18 Same structure, different depth when problem-solving

Here are nine identical squares. Calculate the area of the shaded part.

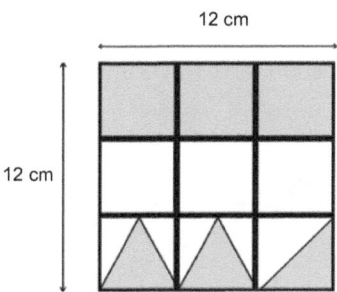

12 cm

12 cm

FIGURE 4.19 Same structure, different depth when problem-solving

We can see that the component parts are important for problem-solving and it is the children's ability to make connections with these component parts that enables them to solve problems.

OK, over to you. Here is a problem; what are the component parts of this problem? What facts do children need to recall in order to solve it? How would you approach it? What would a good 'outcome' or 'finished task' look like in children's books?

Six identical right-angled triangles are arranged to make a rectangle. Calculate the missing length.

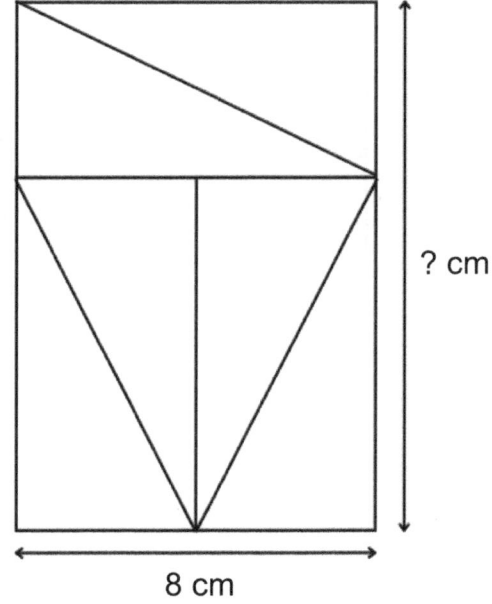

? cm

8 cm

FIGURE 4.20 Same structure, different depth when problem-solving

This problem requires children to recognise the square in the rectangle first and to identify the fact that the sides of a square have equal lengths. This is shown in model 1 in Figure 4.21. Children then use their knowledge of the whole length of the base of two triangles and divide that by 2 (8 cm ÷ 2 = 4 cm) and then connect that with the base of the other triangle at the top of the rectangle (recognising that it is orientated in a different way, see model 2). Fluency and recall of multiplication tables facts and related division facts would be useful for this problem.

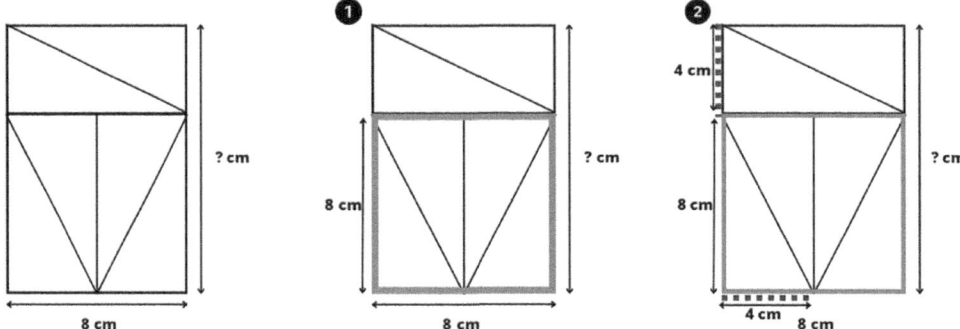

FIGURE 4.21 Showing the component parts in order to solve a problem

I like this problem. It allows children to make connections and think mathematically by linking the smaller component parts to solve a complex problem. It requires creative mathematical thinking. But I also like it because it can be developed even further. We might ask children to calculate the area of each of the triangles or the area of the shaded parts.

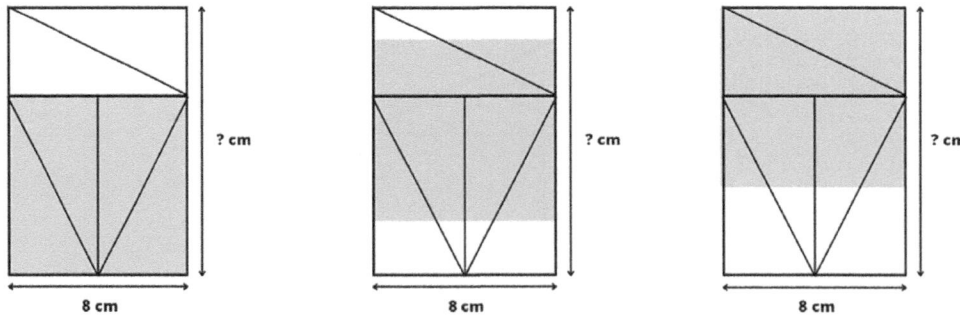

FIGURE 4.22 Same structure, different depth

Or, as in the example in Figure 4.23, we may ask for the perimeter of the pentagon. Or we may ask for the perimeter and area of the shaded square.

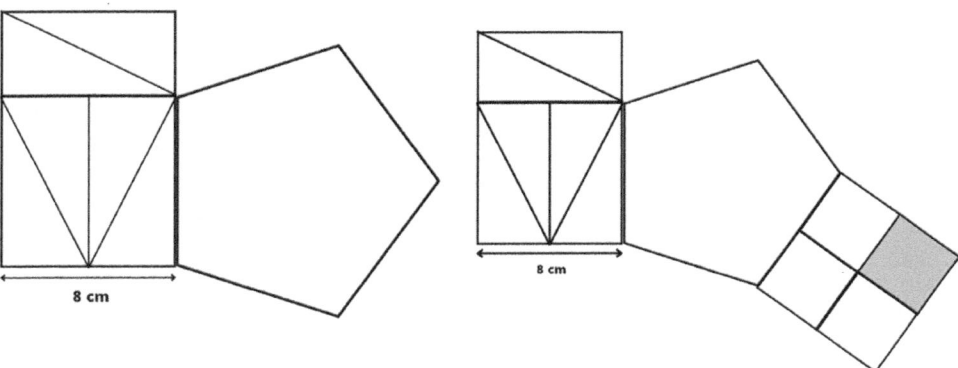

FIGURE 4.23 Developing thinking through slight changes, allowing children to connect their thinking

These problems would be part of a wider lesson on either area or perimeter. They'd be tightly linked to what the children were learning. They may even be turned into a goal-free problem (more on these in a moment).

Faded examples

Essentially, when we teach problem-solving, we are offering a worked example by breaking down the component parts. Some children do not know where to begin and there may be too many component parts to manage. This can often be the case for some of the questions on the end of Key Stage 2 SATs papers. Faded examples offer a strategy that includes a worked example but then applies variation theory to keep the structure the same so that children can work through examples. Each example is different and removes one step of the worked example. Basically, the scaffolding is removed step by step until children complete a problem without any scaffolding support at all. This reduces the cognitive load of the component parts and allows children to think through the order and structure of questions. It allows children to focus on the component parts and gradually build up to combining them to solve problems.

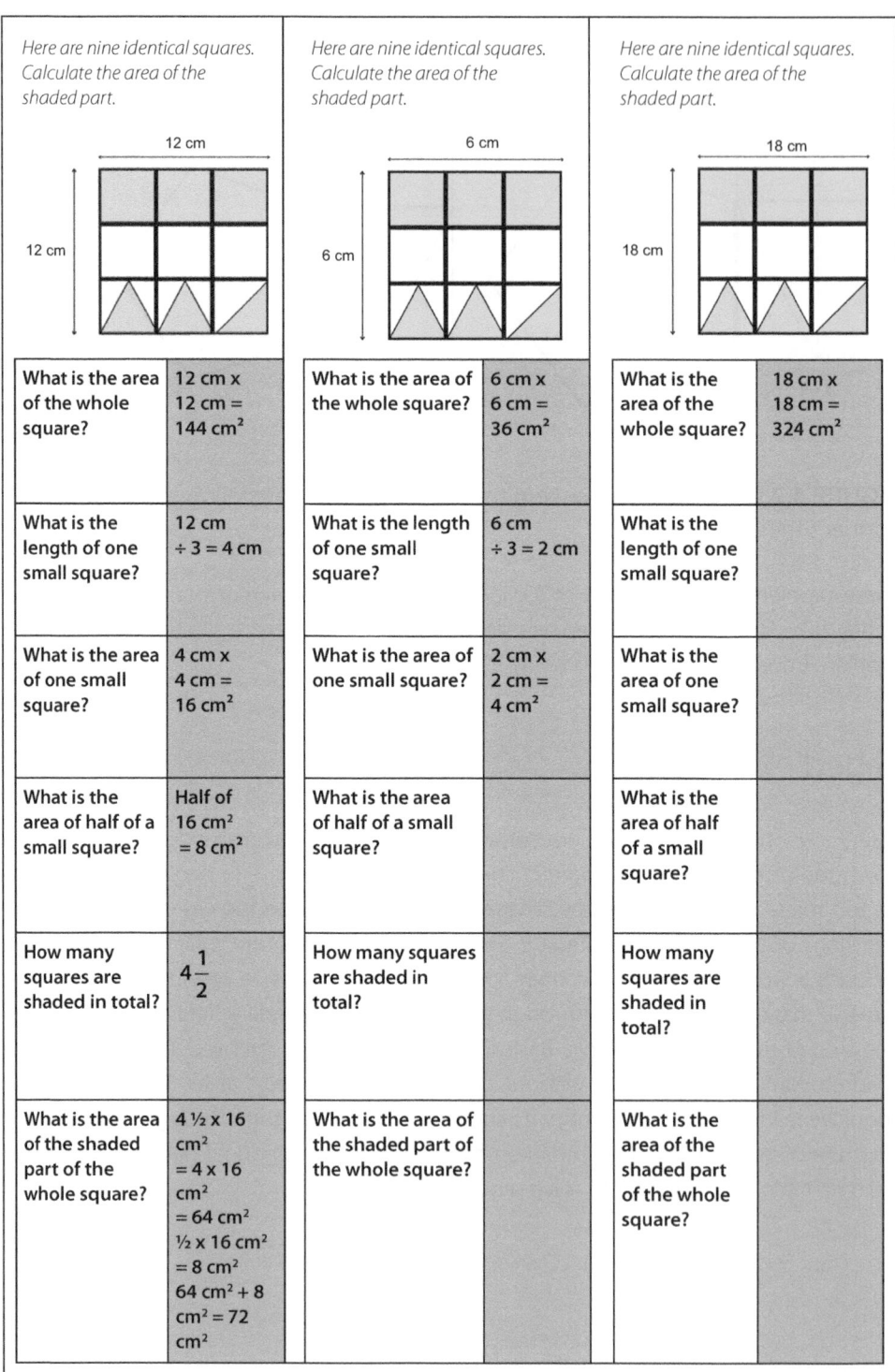

Here are nine identical squares. Calculate the area of the shaded part.		Here are nine identical squares. Calculate the area of the shaded part.		Here are nine identical squares. Calculate the area of the shaded part.	
12 cm / **12 cm**		**6 cm** / **6 cm**		**18 cm** / **18 cm**	
What is the area of the whole square?	12 cm x 12 cm = 144 cm²	What is the area of the whole square?	6 cm x 6 cm = 36 cm²	What is the area of the whole square?	18 cm x 18 cm = 324 cm²
What is the length of one small square?	12 cm ÷ 3 = 4 cm	What is the length of one small square?	6 cm ÷ 3 = 2 cm	What is the length of one small square?	
What is the area of one small square?	4 cm x 4 cm = 16 cm²	What is the area of one small square?	2 cm x 2 cm = 4 cm²	What is the area of one small square?	
What is the area of half of a small square?	Half of 16 cm² = 8 cm²	What is the area of half of a small square?		What is the area of half of a small square?	
How many squares are shaded in total?	$4\frac{1}{2}$	How many squares are shaded in total?		How many squares are shaded in total?	
What is the area of the shaded part of the whole square?	4 ½ x 16 cm² = 4 x 16 cm² = 64 cm² ½ x 16 cm² = 8 cm² 64 cm² + 8 cm² = 72 cm²	What is the area of the shaded part of the whole square?		What is the area of the shaded part of the whole square?	

FIGURE 4.24 Faded examples involving area

Goal-free problems

As we have seen, interesting problems lead us to talk and think. A good question or task will require children to think, hypothesise, generalise, accept, refuse, criticise, wonder, imagine and think deeply. A good question and, importantly, a good answer can be pored upon, proved or disproved. Sometimes, simply presenting a carefully considered representation or calculations with all text removed can be an effective way in which to stimulate genuine reasoning, wonder, imagination and many other creative thoughts, while unlocking prior knowledge of concepts from across the maths curriculum.

Goal-free problems are a form of open-ended maths investigation that do not initially have a specific answer. They are designed to encourage students to think creatively and critically, and to develop their problem-solving skills. Unlike goal-specific questions, which have a single correct answer, goal-free problems require students to apply their knowledge in a problem-solving context. This approach can reduce cognitive overload and develop flexible thinking, allowing children to bring some of their knowledge of maths concepts to the forefront.

The idea of a goal-free problem shows an image or a concept on the screen. Three broad questions can be asked to get children talking: 'What do you see?', 'What do you know?' and 'What could you find out?' These questions allow children to retrieve thinking from prior knowledge and consider how it may link to the concept shown. With each click of the screen, the problem reveals some text that gradually refines children's thinking. In some cases, some of the information becomes redundant and it can be quite humorous for children when the 'reveal' is shown, because they already know this as some of it seems obvious. The 'slow reveal' can reduce cognitive overload. It refines thinking and focuses the mind on one goal-specific problem. Goal-free problems are versatile – they can be developed into almost any maths lesson.

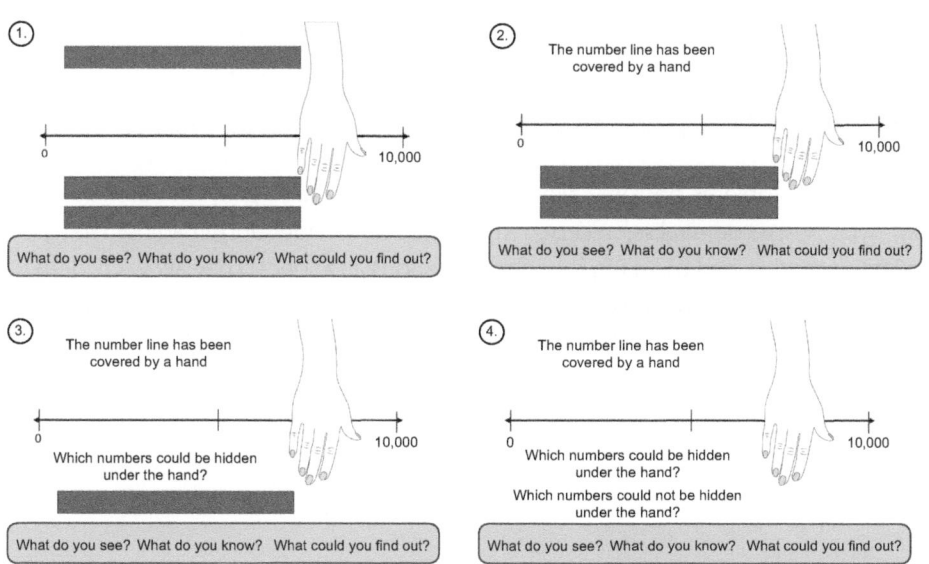

FIGURE 4.25 Goal-free problems slowly reveal criteria to focus thinking

To implement goal-free problems across a school, teachers should first be trained on how to create and use them effectively. They should be encouraged to design goal-free problems that are challenging yet accessible to all students. Once teachers have developed their own goal-free problems, they can begin to incorporate them into their lessons. The value of teaching in this way lies in promoting critical thinking and problem-solving skills. By encouraging students to think creatively and critically, goal-free problems help students to develop a deeper understanding of mathematical concepts. They also help students to develop their problem-solving skills, which are essential for success in mathematics and other subjects. As ever, the power lies in the pedagogy. Using goal-free problems in isolation may not be as effective as embedding them into well-planned sequences and coherent lessons.

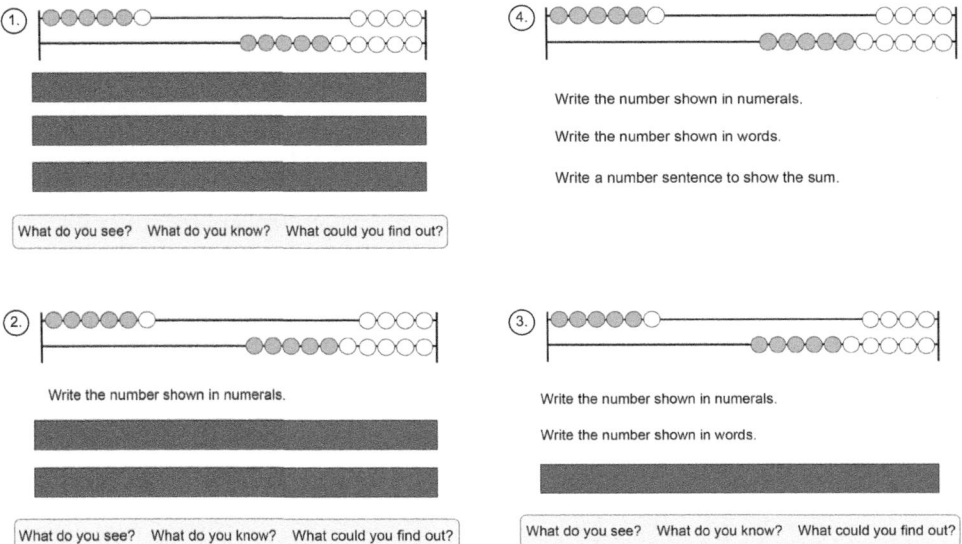

FIGURE 4.26 Goal-free problems slowly reveal criteria to focus thinking

These goal-free problems not only allow for discussion but also allow children to think deeply, hypothesise, generalise, accept, refuse, criticise, wonder and imagine. They also reduce the cognitive load of asking too many questions all at once. They allow children to change their mind part-way through, as the goal becomes increasingly specific. The rich talk and reasoning in lessons can be very powerful. You can track back to the problems that we looked at earlier and consider how we might make those goal-free too!

The power of a visualiser

I mentioned the benefits of the visualiser earlier in this book. I, like many, used to rely on a scheme and ready-to-go presentations and tasks without giving too much thought to *how* I might teach a concept. It was convenient and quick in a busy day in the classroom with lots of other subjects to teach across the day. I often found that while my input had been clear, the output that children completed was often underdeveloped. I spent some time wondering how I might develop this area of the curriculum. How do you show children how to develop their thinking from activities that take 30 seconds to activities that take ten minutes, and which do not involve just giving children more questions but rather require deep and rich thinking? Working on the idea that children learn what they attend to, I started to use a visualiser to model some of the questions that they had in their books. It gave me flexibility within a lesson to model exactly how I would approach a question and how I would break it down in steps if necessary and, more importantly, it allowed me to model my thinking and specific maths vocabulary. In short, I was deliberately teaching children how to reason effectively out loud and in their books. After a few weeks, I started to notice children approaching the tasks that I had set differently. They'd spend more time poring over them, wondering, imagining and thinking deeply. I noticed that children were drawing models (revealing the structure of problems and the maths that they were learning), combined with some sentences alongside the models that demonstrated their reasoning.

The depth of these tasks came into their own when I was able to extend children in different ways to deepen their thinking, to show the structure in a different way or to reason deeply. Over time, and once children were efficient at producing mathematical models and drawings in their book, children could be more independent. As the year progressed, I was always struck by their creativity and links to other areas of the curriculum.

Subject knowledge

To help children to think and develop mathematical thinking, teachers need good subject knowledge. Without good subject knowledge of what we are teaching, we can rely too much on how we were taught when we were at school or how our parents taught us. I remember asking a child in Key Stage 1 about a lesson they had had (which was meant to be) on inequalities. The child told me with a huge smile on their face that crocodiles eat the biggest numbers. When I asked them whether crocodiles actually eat numbers, they nodded and skipped off to have their lunch (perhaps to select their favourite numbers to eat too). The point that I'm making here is that the child had no idea about the important mathematical structure of inequality (something that they would use way beyond primary school). Instead, they'd been taught a short-cut – a trick and gimmick that related to nothing at all. I can see the well-intentioned 'fun' that is sitting behind this approach, but I know that when children leave the lesson, they think of nothing other than crocodiles, which misses the point of the inequality symbols completely. And how do we go about teaching the idea of equivalence with crocodiles? Answers on a postcard please…

FIGURE 4.27 How we teach and represent informs how children think – inequality symbols have no connections to crocodiles at all

There may be two things happening here. Subject knowledge may be the issue (and this gimmick might support the adult teaching this) or it could be a pedagogical issue. Making lessons superficially 'fun' is often a distraction and skews the point of a lesson. Context is fine, but gimmicks and tenuous links should be avoided. Some teachers may teach children; others may teach maths. We want our teachers to teach children maths by blending good subject knowledge with their effective pedagogical approaches. Being a subject expert in primary schools can be difficult to achieve. Maths is one part of a whole primary curriculum. We know that subject knowledge and pedagogical knowledge can develop over time, but it does not do so automatically and it cannot be left to chance. Support and challenge from other teachers are necessary. Nothing is more fun than arriving in a maths lesson and really understanding something. The representations in Figure 4.28 show the power of representation using the inequality symbols. We can see how children can develop conceptual understanding that $1 < 3$ and $3 = 3$ and $3 > 1$.

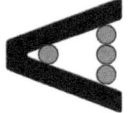  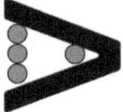

FIGURE 4.28 We might use manipulatives and lollipop sticks to demonstrate the inequality and equals signs effectively

Maths is full of symbols that link directly to vocabulary, concepts and facts. We should teach them with due regard. Now let's deepen this structure and really think about it. How could we get children who grasp this to think even more deeply? Perhaps number composition could be a way in which to strengthen thinking. What do the representations in Figure 4.29 show?

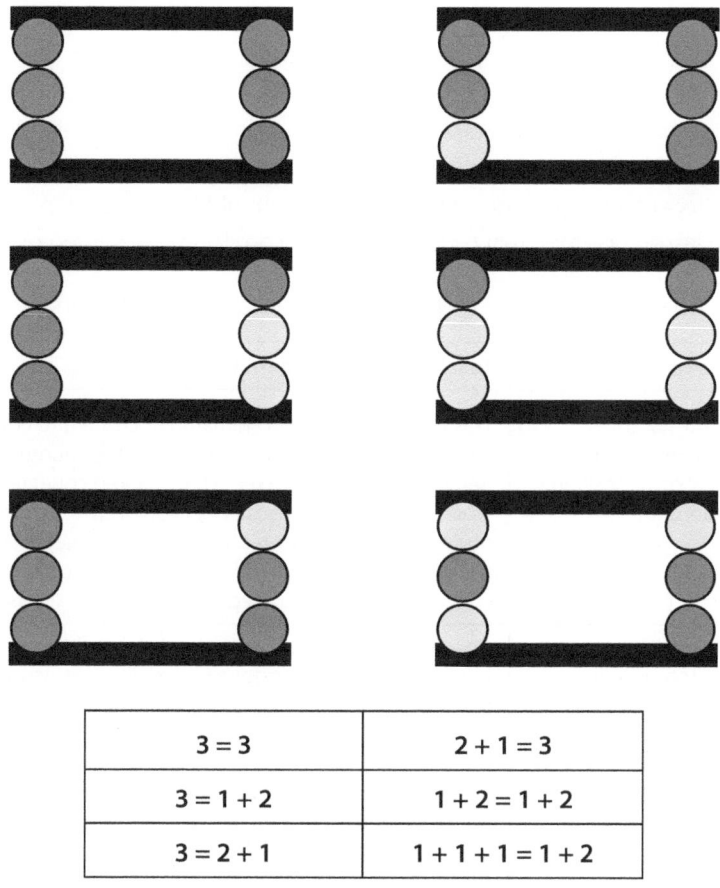

3 = 3	2 + 1 = 3
3 = 1 + 2	1 + 2 = 1 + 2
3 = 2 + 1	1 + 1 + 1 = 1 + 2

FIGURE 4.29 Showing the usefulness of equivalence combined with composition – this could be achieved using double-sided counters

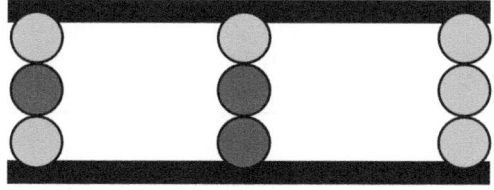

FIGURE 4.30 3 = 3 = 3 and also 1 + 1 + 1 = 1 + 2 = 3

Designing thinking

I often hear 'teaching and learning' said together, as though they are mutually exclusive to each other – if learning has not happened, no teaching has occurred. Teaching and learning are not necessarily synonymous. I prefer just 'teaching'.

Teaching can lead to learning (but not always). If you have done something in the classroom and the children have not learned it, then no teaching has taken place. So, how do we design lessons that develop mathematical thinking? Table 4.1 describes a lesson that I taught recently that was designed (as they all should be) to develop mathematical thinking. The lesson was for a Year 6 class on multiplying decimals by integers.

TABLE 4.1 A Year 6 lesson on multiplying decimals with integers

Focus/resource	Teaching children to think
Animated tenths and hundredths, revealing place value counters on a tens frame with each click of the screen	This was done using an animated document to count forwards and backwards with place value counters on a tens frame. It allowed me to see what children could remember while strengthening their understanding with flexible questioning: 'What is 1 tenth less than 5 tens?' or 'What is double one hundredth?' This was essential to the lesson that children were about to begin.
Counting forwards and backwards in tenths and hundredths with animated place value counters on a tens frame	Children then counted in tenths and then hundredths out loud with me. This was done in two different ways: 'One tenth… two tenths… three tenths…', before swapping to unitise the language – 'One one tenth, two one tenths, three one tenths…'. This was done using place value counters and a tens frame. When I said 'STOP', children wrote down the corresponding decimal. I then asked what each digit in (for example) 1.7 represented, to draw out the language and thinking.
I displayed the following calculations: 7 x 2 = 14 7 x 0.2 = 1.4	I asked children what they noticed and gave them some time. When children fed back, I (re)taught children that 7 was a factor, 2 was a factor and 14 was the product. I then asked them to discuss again and children arrived at the generalisation: that if you keep one factor the same and divide the other factor by 10, then the product is divided by 10. We then shared other examples with place value counters to represent this structure.

TABLE 4.1 (Continued)

Focus/resource	Teaching children to think
I displayed the following calculations: $3 \times 4 = 12$ $3 \times 0.4 = 1.2$	Two representations were shown on the screen. Children were asked what each digit in the calculations represented. The children reasoned: 'The factor 3 represents the three groups, the factor 4 represents the 4 ones in each group and 12 represents 12 ones as the product.' Children then generalised: 'The 3 represents the three groups, 0.4 represents the four tenths in each group and the 1.2 represents the 12 tenths as the product.' This was repeated for a few more different examples so that children could practise the structure of this.
I displayed the following calculations: $3 \times 2 =$ $3 \times 0.2 =$ $6 \times 2 =$ $6 \times 0.2 =$ $12 \times 2 =$ $12 \times 0.2 =$	Children were then given these examples to work through. At the end of this, I asked what they noticed about the factors this time. They agreed that the first factor was doubling (3, 6, 12). Children were able to then generalise that if one factor doubles and another stays the same, the product doubles. At this point, I asked children to continue with the sequence themselves in their books for a few moments. Children were spotting patterns and explaining their thinking at every stage of the lesson. I then asked, 'What would $1.5 \times 0.2 =$?' Children referred back to their first questions and began to reason about halving a factor and the impact on its product.
9×0.04 ___ 6×0.04 8×0.06 ___ 9×0.06 8×0.03 ___ 9×0.03 8×0.03 ___ 0.03×8 $5 \times 0.03 + 3 \times 0.03$ ___ 6×0.03 $+ 2 \times 0.03$	Next, I presented this sequence of questions to children one by one where they had to compare each expression using <, > or =. I asked them what method they would use. Children could draw upon their previous generalisations and could notice what was staying the same and what was changing in each question. This allowed them to answer quickly (without necessarily arriving at the answer), due to them understanding the structure of the questions.
0.9cm 8.1 cm 9 cm 1.5 cm 5.4 cm 6 cm	Then I presented the following image and asked children 'Which lengths *could* the chain be, and which lengths could it *not* be? This prompted the children to reason about the powers of ten and then apply it to the problem. I then asked children to design their own question based on this, showing the lengths that it could be and the lengths that it could not be.
$3.3 \text{ g} \times 3$	Next I posed a contextual problem that could be solved mentally but multiplying the tenths and the ones without any exchanges. I represented this using place value counters and combined them to show the total. This was done using a place value grid, in preparation for the formal written method.

(*Continued*)

TABLE 4.1 (Continued)

Focus/resource	Teaching children to think
3.3 g x 4	After that, I modelled this again with the place value counters and a place value grid. I modelled the exchange and showed children that 10 tenths are equivalent to 1 one, and so we need to exchange the 12 tenths for 1 one and 2 tenths. Alongside this, I modelled the formal written method at each step.
2.4 x 2 2.44 x 2 2.42 x 2 4.2 x 2 42.2 x 4 0.24 x 4 0.42 x 4	I then gave children some time to practise using the formal written method in their books. Children still had access to the place value counters and place value grid to use if they needed to. The first few examples in the sequence I gave children allowed them to calculate mentally. We discussed how it would not be necessary to use a formal method for some (if not all) of the questions. I wanted the children to gain procedural fluency with the written method, before the next lesson moved on to less routine questions.
12 x 2 1.2 x 2 0.24 x 2 4.8 x 2 0.12 x 2	To finish the lesson, I asked children to study the calculations and to rank them in order from easiest to most difficult. I reminded children here of the key vocabulary of factor, product, tenths, hundredths and place value. I wanted them to think of the methods that they would use and the process that they would have gone through to order them. Some used the first calculation to help them with the other calculations, thus making connections and links in maths.

Obviously, it's difficult to capture the actual thinking that went on in the lesson, and this is simply my interpretation of how I designed the lesson to inspire thinking.

But what about greater depth? 'Greater depth' is not a label for those who can do maths. Those children who are working at a greater depth of understanding are those children who have mastered the component parts of much of the curriculum and can use them in a variety of different contexts. In other words, they make connections across the curriculum. Just as we saw with the use of number lines in Chapter 3, using these in the context of mass, weight or capacity allows children to make sense of the maths that they are using. They have the same 'ability' as any other child in your classroom but attain at a deeper level because of their depth of understanding.

Making connections

A huge focus – if not the main focus – of teaching for mastery is for children to make connections across the curriculum. To do this, children need to be secure with the fundamentals – the components of the curriculum. Let me ask you how your (Key Stage 2) class might solve: $1 - \frac{3}{4}$. Perhaps you could ask them what they would do tomorrow. I wonder how many are wondering whether children would convert, draw number lines, use bar models, etc. Now, what would you hope that they would do to solve this efficiently? What teaching has had to happen in order to do this? Here, we want children to make the connection between 1 always being equivalent to a fraction that has equivalence when its numerator and denominator are the same. We need to teach children the generalisation that: $1 = \frac{2}{2}, \frac{3}{3}, \frac{4}{4}, \frac{5}{5}, \frac{6}{6}$ and so on. Now children can make connections when subtracting: $1\left(\frac{4}{4}\right) - \frac{3}{4}$. What's more, children can use this to go on to solve $1 - \frac{5}{6}$ and $1 - \frac{1}{8}$ and begin to apply the rule flexibly to $2 - \frac{1}{8}$.

This can and should be done alongside representations (such as bar models for the example above) and will naturally lead to rich reasoning. Doing fewer examples better and deeper is the trick here. Being clear on the structures that you are teaching and how they link together is of importance. We can then see how we can design questions for children to answer around this:

- $1 - \dfrac{3}{4} =$

- $1 - \dfrac{3}{5} =$

- $1 - \dfrac{3}{6} =$

- $2 - \dfrac{3}{6} =$

- $3 - \dfrac{1}{2} =$

Notice the careful choice of examples chosen above (more on this in Chapter 6). Here, there's no reliance on long-winded conversions or any other short-cut or gimmick. In designing sessions to include mathematical thinking at every point of a lesson and series of lessons, we're paving the way for children to look for patterns and relationships, make connections, conjecture, reason and generalise. In other words, they're learning to think like mathematicians. Maths can become exciting and genuinely intriguing for many.

We teach children to count in multiples of 2, 5, 10 (and all other times tables), as well as 25, 50, 100 and 1,000. But why? Where is the link? How does this connect to other areas of the curriculum? What springs to mind first? For me, teaching these multiples links to statistics. Often, charts and graphs can increase or decrease in multiples of 25, 50, 100 or 1,000. Now, we might teach these at the beginning of the year with place value or number facts, but it is important to know how they connect to areas of our curriculum so that we can maximise their impact. If children forget, don't worry. We can reteach and remind, recap and retrieve, because to learn means to remember, build and apply. Having these number facts secure is essential later down the learning journey. It is for this reason that careful and deep thinking about your curriculum design, and where and when concepts and skills are taught and revisited is of paramount importance. These are the building blocks that will unlock the curriculum in all its interconnected beauty!

Personalisation rather than differentiation

We looked at different representations in Chapter 3. They can help to support how children think. Their beauty lies in the structure that they reveal. We want children to think deeply about mathematical structures. The National Curriculum requires children to solve routine and non-routine problems. But how do we do that? I have written at some length about the damaging effect that traditional differentiation and three (or more) ways of differentiation can have on children. Instead, I might present most or all of the children in my class with a problem and talk through it. This becomes a worked example and is perhaps the starting point for an 'I do, we do, you do' approach to teaching and learning. Let's take the example of a triangle in Figure 4.31. Imagine that you have taught children about the degrees in a triangle. Routine problems would require children to remember that there are 180° in a triangle and then calculate the missing angle. How could you develop this problem to a different depth, so that it becomes non-routine and perhaps requires children to make connections across the maths curriculum?

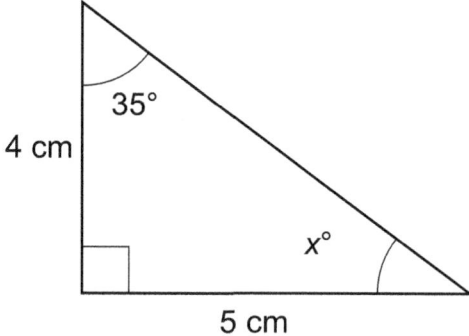

FIGURE 4.31 A problem that could be shared with the whole class

Here is one suggestion. When I first started to consider how to get children to think, I actually started giving children fewer paper tasks or worksheets. Instead, I would use one well-chosen example with all children, and as I gained confidence in embedding this, I would often find myself working with groups and drawing images or questions into their books to personalise their thinking. After some time, wonderful things started to happen. Children were making connections and making their own suggestions as to how problems could be approached and solved. Children were thinking in flexible ways and often had novel approaches to thinking about problems. We might even be able to see the subtle changes in measurements and how they connect to each other. This may allow our questioning to steer towards the area of triangles. Once children see the patterns and connections, they might have a go at drawing what the next problem may look like. We can really push thinking by asking children what the mean area of all the triangles is. This way of working can take some time, but the opportunities for thinking that it presents are insurmountable. Figure 4.32 is an example of where I might start with all children and how I might personalise the tasks in the lesson.

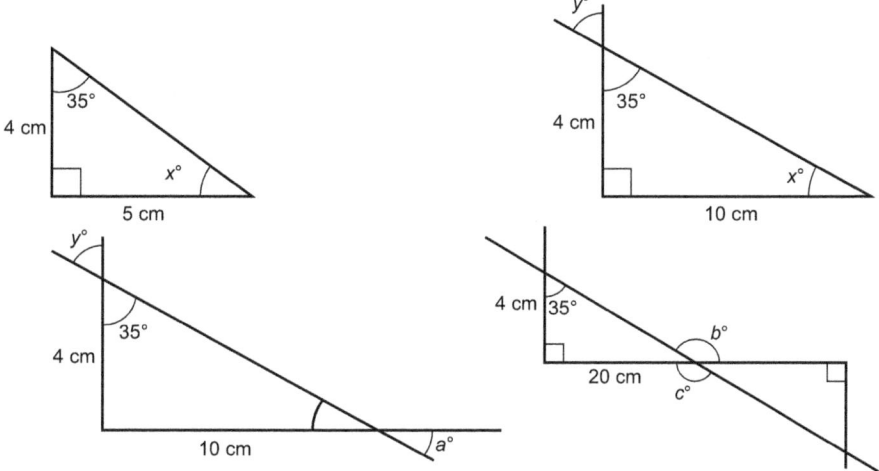

FIGURE 4.32 An idea of how problems can be designed and redesigned to deepen thinking

Let us try this again in the context of decimal fractions. Imagine that we have taught children that $0.01 = \frac{1}{100}$ and other related facts, alongside 10 tenths = 1 one and 10 hundredths = 1 tenth. How might we keep this structure but ensure that children can go to great depths of understanding in one lesson? How can we personalise this for learners to explore the structure? Take a few moments to consider or perhaps write down a sequence of questions that children might complete in their books.

Here is one suggestion of how we might get all children to think deeply:

- $0.01 = \dfrac{1}{100}$

- $0.02 = \dfrac{}{100}$

- $0.03 = \dfrac{1}{100} + \dfrac{}{100}$

- $0.1 = \dfrac{}{100}$

- $0.11 = \dfrac{11}{100}$

- $0.22 = \dfrac{22}{100}$

- $0.22 = \dfrac{}{10} + \dfrac{}{100}$

- $\dfrac{}{} \cdot \dfrac{}{} = \dfrac{4}{10} + \dfrac{4}{100}$

- $\dfrac{2}{10} + \dfrac{3}{100} > \dfrac{3}{100} + \dfrac{}{10}$

- $1 = \dfrac{5}{10} + \dfrac{}{100}$

I'm sure that the next few pages could be populated with this sequence, maybe going back and adding even smaller steps between questions and then deepening it in different ways. Perhaps introducing multiplication or making links to percentages would be appropriate in some classes when deepening this structure. Carefully designed tasks allow children to think with increasing creativity and flexibility. I'd love to see and hear about some of the approaches that you have tried to implement in your classrooms. Please do message them to me on social media (@mrbeeteach).

Retrieval practice

There has been a recent shift with regard to retrieval practice. Lots of schools have read research from Rosenshine and considered the forgetting curve (Ebbinghaus, 1885). Many lessons in schools now begin with well-intentioned retrieval. In my experience, this is done with varying degrees of success. Quite often, retrieval can be a list of a few questions for children to get on with as they enter a classroom. As teacher time is limited, this can often mean that retrieval is random and superficial or left to whatever a scheme happens to have suggested on a particular day.

We must consider what we want children to know and remember. Trying to remember a whole curriculum can be daunting, and decisions on what to include, in busy schools, are often quick. This means that children are busy doing stuff but what if they struggle and cannot remember? Do we place children on the wrong foot at the very start of a lesson by presenting a list of questions that they cannot do?

Knowing what to prioritise for retrieval might be a good starting point. Do we want children to have more practice on particular areas of the maths curriculum? If so, why? What is the rationale behind this? Many published schemes do this organically, devoting more time to place value, the four operations, fractions and shape, and less time to subjects such as algebra or statistics. Consider the following now: What are your curriculum promises to the teacher next year? What are you going to relentlessly focus on at the start of each lesson? What is the essential knowledge that children need to access the wider maths curriculum? For teachers in Year 2, this might be a focus on the criteria in the 'progression in number facts' document (more of this in Chapter 5). This could make your retrieval practice even tighter and more focused. If children spend full terms or a full academic year focused on those skills, when we come to teach them as part of our main curriculum, there is a good chance that children will arrive well-prepared for success because of your deliberate choice on what to retrieve.

Of course, this can not be completely rigid. AFL should inform the decisions that we make as teachers. If we know that children still need practice with, for example, doubles in Key Stage 1, it makes good sense to include these as part of retrieval until children achieve automaticity. The trouble is that if we display a double on the board each morning as part of our retrieval and children are still not getting it, what do we do then? Here, I think that we need a hybrid of getting on and doing at the start of a lesson combined with retrieval as directly explicit (re)teaching.

At different points of the school year, it will be appropriate to focus on different things. For example, in September, we may go back to priority skills covered in the previous academic year to remind children after a long summer break before building on new learning in their current year group. As the year progresses, the teacher has their priority areas for retrieval but is making informed decisions based on AFL about which needs greater focus each day.

Times tables are integral to accessing mathematical ideas. If children are fluent, they will probably be doing more challenging types of mathematics because they need to work out the times tables fact before applying it to a problem. For example, when calculating the area of compound shapes (as we saw earlier) or converting fractions, this requires quick recall in order to apply the times tables facts to a problem to solve. Prioritising times tables and explicitly teaching them as part of retrieval is always going to be time well spent.

This approach does come with a word of warning. We must be careful not to narrow the curriculum for children. Shape is important and so are statistics. But there is a reason why we might prioritise counting in multiples of 2, 3, 5, 10, 25, 50, 100 and 1,000, as these tend to be intervals on scales when teaching statistics. We might want to consider what the priority is around shape. What do we want children to know and revisit or to (re)teach at the start of some of our lessons? What is the most important knowledge for children so that they do not arrive in Year 6 with a blurred understanding of the properties of a square or hexagon, for example? Again, this is something that could be done collaboratively as a school team.

Sitting with the curriculum for geometry from EYFS to Year 6 may be a good starting point, before committing the areas on which you want to focus across each year group. Threading this through your retrieval can have a significant impact. Committing this to a progression document or attaching it to the 'number facts document' to include a geometry focus will allow your teachers to 'track back' and revisit previously taught concepts.

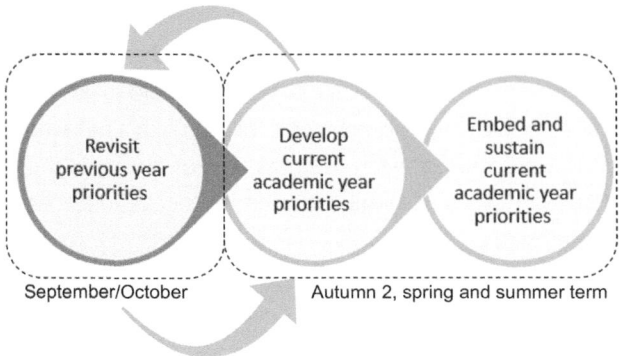

FIGURE 4.33 Planning a progression document

This deliberate and systematic approach to retrieval will ensure that more children make progress in your curriculum.

Key takeaways

In this chapter on mathematical thinking, we have delved into the importance of fostering a deep, conceptual understanding of mathematics in students. The central premise is that mathematical thinking is not merely about performing calculations or following procedures, but involves a holistic understanding that integrates fluency, reasoning and problem-solving into a cohesive learning experience. This approach ensures that students do not just learn mathematics but think mathematically.

Emphasising the interconnected nature of mathematical fluency, reasoning and problem-solving is imperative. The pitfalls of treating components as separate entities can lead to a fragmented understanding of mathematics. Instead, I advocate for an integrated approach, where fluency supports reasoning and both are essential for effective problem-solving.

By breaking down complex problems into manageable parts, students can tackle them step by step, building confidence and competence. This method reduces cognitive load and helps students to see the connections between different mathematical concepts. For instance, understanding the properties of shapes and the relationships between their dimensions can simplify seemingly complex geometry problems.

Worked examples and faded examples are effective strategies to scaffold learning and to teach children how to think mathematically. Worked examples provide a clear demonstration of how to approach a problem, while faded examples gradually remove support, allowing students to build independence. This method is particularly useful in helping students to internalise the problem-solving process and apply it to new situations.

We have also explored the importance of creating genuine and engaging problems that stimulate students' curiosity and encourage deeper thinking. Tasks should be about not only finding the correct answer but also understanding why that answer is correct. This involves encouraging students to hypothesise, conjecture and justify their reasoning, which cultivates a more profound mathematical understanding and appreciation.

As we transition to the next chapter on mathematical fluency, we will build on these foundational ideas. As we will see, mathematical fluency goes beyond mere speed and accuracy; it encompasses a deep understanding of numbers and operations that allows students to manipulate them flexibly and confidently. We will explore strategies to develop this fluency, ensuring that students can apply their mathematical knowledge effectively in a variety of contexts.

By integrating the insights from this chapter on mathematical thinking with the upcoming focus on fluency, I aim to provide a comprehensive framework that supports students in becoming not only proficient but also insightful and creative mathematicians.

5 Fluency
By Emily Pringle

There are so many elements to great mathematics teaching, but getting our approach to fluency right is perhaps one of the most important of all. A child's success with mathematics can stand or fall on the basis of the opportunities that they have been given to develop fluency with facts, methods and strategies. Both in the past ten years of teaching maths in primary schools and, more recently, in my work as a primary adviser, developing fluency has remained at the forefront of my priorities. So why are the stakes so high? Why is developing fluency so important?

Why prioritise factual fluency?

Let us consider two pupils. The first can securely recall all of their times table facts up to 12 x 12. The second has limited recall of the same core facts. Imagine that both children are asked to solve the problem in Figure 5.1.

Calculate the area of the following rectilinear shape.

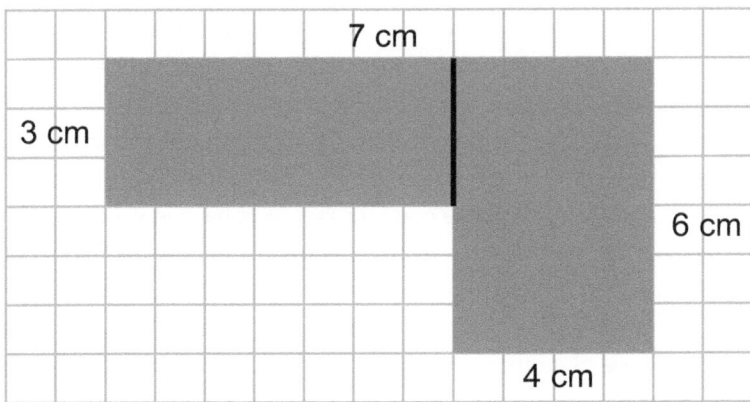

FIGURE 5.1 What fluency would children need to access this type of problem?

Having recently studied area in their Year 4 class, both children have a secure conceptual understanding of how to calculate the area of rectilinear shapes. Both understand that you find the area of a rectangle by multiplying the length by the width. Both understand that, when finding the area of rectilinear shapes, the shape must first be split into rectangles. Using this foundational knowledge, there are multiple ways of solving this problem. But let us imagine that both children choose to break up the rectilinear shape into two rectangles as illustrated above. In order to successfully calculate the total area, both children need to be able to calculate the core facts of 3 x 7 and 6 x 4. The first child can do this quickly and efficiently, mentally recalling that 3 x 7 = 21 and 6 x 4 = 24 and combining them to reach a total of 45 cm². However, the second child is not yet fluent with these facts. Instead, it takes them a while to calculate 3 x 7 and 6 x 4, as they skip-count in 3s (3, 6, 9, 12, 15, 18, 21) and then 4s (4, 8, 12, 16, 20, 24). Not only does this method take longer but, given both the inefficiency and the increased number of steps involved, the child is also more prone to make errors in their calculation. As a result, the second child is much less likely to arrive at a correct answer to the calculation. Here are two children, working on the same problem, yet one is forced to do more work than the other to get to a correct answer. It is their limited knowledge of their times table facts that is holding them back. This is what Gray and Tall (1978) are referring to in their research when they say that some children are forced to do a more difficult type of maths than others. We are making the maths harder for the children who need the most support.

Addition and subtraction fluency

The same is true for addition and subtraction fluency. When teaching column addition and subtraction in Year 4, I used to find that the children who struggled the most were frequently those who were not yet secure with their addition and subtraction facts within 20. Often, these children would resort to inefficient counting strategies to calculate 9 + 8 or 15 – 6, rather than being able to simply recall the facts or use quick mental strategies. If we want to give all children fair access to the mathematics curriculum, we need to have systems in place that address this problem. If we are not doing something to address the barrier that a lack of number fact fluency causes, then the gap between our lowest and highest attaining pupils is likely to widen. While access to the curriculum content may be the same, children who are not fluent with these core number facts have fewer and poorer tools at their disposal with which to handle the demands of more complex mathematical problems.

The Ofsted subject report for maths (Ofsted, 2023) discusses the importance of number fact fluency. It is suggested that one of the difficulties is that pupils in Year 3 and 4 can still 'get by' with the use of inefficient methods, such as counting on fingers, to calculate addition facts. However, by Year 6, as the demands of the curriculum increase, these difficulties become much more significant. As such, many difficulties with number fact fluency are not being addressed until the point at which they are causing significant problems. So, what should we do about this? Well, as the report suggests, we need to look at how these facts are acquired in the first place, alongside making sure that we are intervening as early as possible to support those who do not have security with these facts. This approach can be summarised as follows:

1. Ensure that we have robust systems in place for securing fluency with core number facts.

2. Use assessment to carefully identify those who are secure with these facts and those who are not.

3. Intervene early to provide high-quality additional support for those who are not yet secure with these core facts.

What makes pupils fluent?

In order to develop robust systems for securing factual fluency, we firstly need to consider what we mean by fluency and precisely what it is that we need to be fluent with.

So, what do we mean by fluency? Well, the first thing to note is that, in maths, fluency and understanding are intrinsically linked. You can't have true fluency without deep understanding. Nor can you have deep understanding of a concept without having some level of fluency. According to the NCETM:

'Procedural fluency and conceptual understanding are developed in tandem because each supports the development of the other.' (NCETM, 2022)

To illustrate why both are necessary, let us consider the following example:

64 bottles of lemonade are packed into 8 boxes. How many bottles are in each box?

To accurately solve this problem, pupils need both understanding and fluency. A child who has secure understanding would be able to recognise, for instance, which operation is needed to solve this problem. The language of sharing should be an immediate indication to the pupil that this problem has a multiplicative structure rather than an additive structure. If lemonade is packed into 8 boxes, then we are talking about '8 lots of __'.

Further, to know that this is a division problem requires understanding that multiplication and division also have different structures. In multiplication, we are given the factors (the parts) and use these to calculate the product (the whole).

8 x 8 = ?

FIGURE 5.2 A problem with a multiplicative structure

However, in division, we are given the product (the whole) alongside one of the factors (a part) and use these to calculate the other factor.

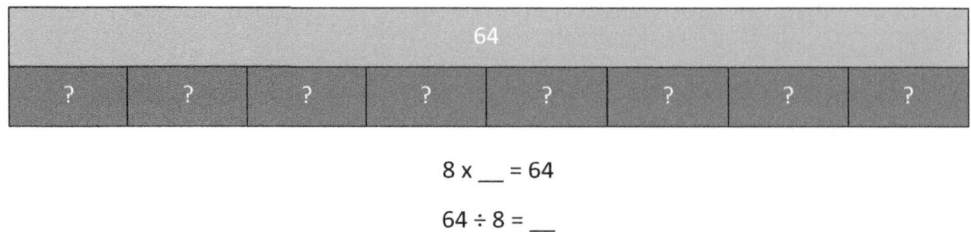

$$8 \times \underline{} = 64$$

$$64 \div 8 = \underline{}$$

FIGURE 5.3 A problem with a division structure

However, understanding that this is a division problem is not enough in and of itself to successfully obtain an accurate answer to the calculation. The pupil also needs an accurate and efficient method with which to calculate $64 \div 8$. Consider the potential ways of solving this division calculation shown in Figure 5.4.

Method 1	Method 2
Dividing 64 into 8 equal groups:	Repeated addition:
	$8 + 8 + 8 + 8 + 8 + 8 + 8 + 8 = 64$
Method 3	**Method 4**
Division by partitioning: 64 40 24 $40 \div 8 = 5$ $24 \div 8 = 3$	Using knowledge of times tables: $8 \times \underline{} = 64$ $8 \times 8 = 64$

FIGURE 5.4 Potential methods of solving $64 \div 8$

Methods 1 and 2 certainly serve a purpose in building understanding of the concept of division. When children are first introduced to division, they will often explore these sharing and grouping methods. However, with larger numbers, these methods become long, cumbersome and very inefficient. Further, due to the number of steps involved, children are much more prone to make an error at some stage. Hence, neither method can be considered particularly effective in this context. On the other hand, Methods 3 and 4 are more efficient approaches. They involve fewer steps and, because there are fewer places to make a mistake, the approaches are also more accurate. As we can see here, there is often a direct connection between efficiency and accuracy in mathematics. An efficient strategy is one that involves using a method with fewer steps. Efficient strategies are often the most accurate – that is, provided that we have the necessary skills to carry out the strategy in the way in which it is intended.

So, our example shows that to be fluent requires both a deep understanding of concepts and the application of efficient and accurate methods. Yet fluency also requires flexibility. The efficiency and accuracy of methods are not static but fluid. What is an efficient and accurate method in one context may not be efficient and accurate in another context.

Let's compare the earlier calculation to another calculation:

1. $64 \div 8 =$
2. $2{,}608 \div 8 =$

For $64 \div 8$, the use of known facts, as illustrated by Method 4, is more efficient than Method 3. If I have secure recall of my multiplication facts, I should immediately be able to use the known fact '8 x 8 = 64' to solve this calculation. However, for $2{,}608 \div 8$, the use of known facts may not be the most efficient and accurate strategy. I would be unlikely to be able to automatically recall '8 x 326 = 2,608'! Therefore, in this case, use of short division is likely to be a more efficient and accurate approach. I could, of course, still use known facts, drawing upon my knowledge of '300 x 8 = 2,400' as a starting point. But, when working with a larger dividend (2,608), this will not necessarily always be the easiest strategy – hence the need for flexibility. To use an efficient strategy, I must first consider the context of the problem, before deciding which method to select. This reinforces the importance of pupils developing and exploring multiple methods for solving calculations. If a pupil learns short division as the sole method for solving division problems, then they will not have the flexibility necessary to select a different strategy when the calculation changes. They are limited in the repertoire of tools that they have to hand to tackle the problem.

So, to return to our original question, what is fluency? As we can see, fluency requires understanding of both what a concept is and how to perform the necessary operations. The above examples also support Russell's definition of mathematical fluency as that which requires efficiency, accuracy and flexibility (Russell, 2007).

What do we need to be fluent with?

Mathematics is made up of different types of knowledge. There are four key categories that all play an important role in the development of fluency.

- **Knowledge of facts:** e.g. I know that there are 90° in a right angle.
- **Knowledge of concepts:** e.g. I know why angles in a triangle add up to 180°.
- **Knowledge of methods:** e.g. I know how to measure the size of an angle using a protractor.
- **Knowledge of strategies:** e.g. I know when to use particular facts or methods.

These types of knowledge are discussed in the Ofsted research review, which also states that: 'Pupils need to systematically acquire core mathematical facts, concepts, methods and strategies to be able to experience success when problem-solving and in order to become proficient mathematicians.' (Ofsted, 2021)

So, if all four types of knowledge are important for developing fluency, how are they connected?

Knowledge of concepts

As outlined earlier in this book, mastery of mathematics is underpinned by the development of a clear understanding of concepts. Learning facts in isolation only leads to superficial understanding. I may be able to recall the fact that there are 90° in a right angle, but without knowledge of the key concepts that underpin this fact, the fact itself lacks meaning. For instance, in order to have secure understanding of this fact, I would need to understand the concept of a right angle and the concept of a degree as a unit of measure. The same is true for knowledge of methods. Understanding why a particular method works helps to build deep and lasting understanding of this method. To illustrate this point, consider the following example:

$$\frac{3}{8} + \frac{4}{8} = \frac{7}{8}$$

I could teach the rule for adding fractions by instructing pupils to keep the denominator the same and just add the numerators. This gives them the tools that they need to be successful with adding fractions. However, without an understanding of why this method works or how it connects with other concepts that they have learned, pupils are left with only a superficial understanding of this method. Further, when pupils learn other rules for multiplying and dividing fractions, they are more likely to confuse these rules. The rules are not connected to their wider mathematical knowledge, and so they are only reliant upon their memory of the process to determine whether their method is correct.

Several years ago, I had the opportunity to take part in an exchange with teachers in Shanghai. This involved watching and discussing maths lesson design with them, as well as two teachers from Shanghai coming to my own primary school to teach our pupils maths. When teaching fractions to my Year 4 class, I was really struck by the way in which the methods used always fostered depth of understanding. When introducing the concept of non-unit fractions, where the numerator is greater than one, children were encouraged to see that these non-unit fractions are themselves made up of unit fractions. For instance, $\frac{3}{8}$ is made up of three lots of $\frac{1}{8}$. As such, when children moved on to adding fractions, they could draw upon this knowledge to help them to understand why we just add the numerators and we keep the denominator the same. They could see that adding fractions follows the same principle of normal addition that they have been learning about since Reception. This process is known as 'unitising'; we are following the same rules of addition but simply changing the type of unit that we are adding.

For example:

3 elephants + 4 elephants = 7 elephants
3 bananas + 4 bananas = 7 bananas

So also:

3 lots of $(\frac{1}{8})$ + 4 lots of $(\frac{1}{8})$ = 7 lots of $(\frac{1}{8})$

Or:

$$\frac{3}{8} + \frac{4}{8} = \frac{7}{8}$$

By building a clear understanding of concepts, we are removing the need to remember methods in isolation. Instead, we are connecting our method of adding fractions with our already secure understanding of addition. Understanding why a method works strengthens our understanding of how to use this method.

So, a secure knowledge of concepts in turn helps us to build security with our knowledge and retention of facts and methods. These facts and methods are what we need to draw upon in order to select an appropriate strategy (Ofsted, 2023). This could be illustrated as shown in Figure 5.5.

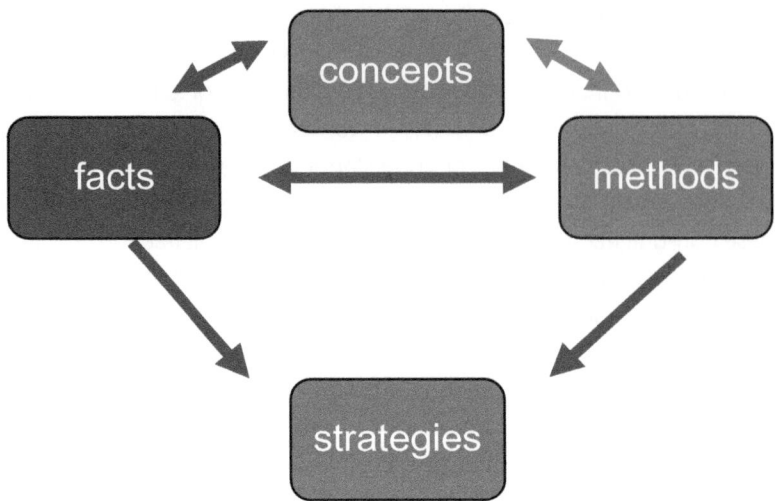

FIGURE 5.5 The connections between concepts, facts, methods and strategies

The bidirectional arrows are used to indicate that this is a two-way process. Knowledge of facts and methods can also help to reinforce our understanding of concepts. For instance, to return to the earlier example, a secure knowledge of addition facts within 20 alongside an understanding of how to add two one-digit numbers in turn make it easier to build our conceptual understanding of the rule for adding fractions and why this rule works. So, our knowledge of facts and methods reinforces our understanding of concepts, in the same way as our knowledge of concepts reinforces our knowledge and recall of facts and methods.

Having established the crucial role of conceptual understanding in building and maintaining fluency with facts and methods, let's move on to consider the three other types of knowledge in turn. How do we build fluency with facts, methods and strategies?

Knowledge of facts and methods

As explained above, it is important that mathematical facts are taught in context. It is important to ensure that they are not just learned in isolation but are embedded as part of our wider web of mathematical knowledge, where meaningful connections are made between concepts.

However, in order for this knowledge to be retained, it is also important that we plan regular opportunities for these maths facts to be consolidated and committed to our long-term memories.

Ebbinghaus's forgetting curve (1885) tells us that much of what is initially retained is lost if this knowledge is not revisited (see Figure 1.3 on page 15 for a visual). This, of course, makes for depressing reading as a teacher – one day after the lesson has finished and maybe only half of that knowledge has been retained. Thirty days later and most of it has gone completely!

So, what can be done to combat this? Well, it certainly highlights the need for the consolidation of knowledge. There is an immediate need for consolidation within the lesson itself. As I'm sure that many teachers can identify with, children do not simply need to hear things once for the message to sink in. Repetition is key. And if children are unlikely to retain all of what is taught, then there needs to be careful consideration of what we want pupils to retain from each lesson. What are the most important bits of knowledge? What are the building blocks that will be needed for future learning? These are the things we must keep coming back to time and time again. Some of those key building blocks come in the form of our maths facts.

The maths facts that I use in the classroom seem to come in three distinct categories, as shown in Table 5.1.

TABLE 5.1 Types of fact with examples

Type of fact	Examples
Number facts	$3 \times 4 = 12$ $9 + 6 = 15$
Mathematical statements or equivalences	January has 31 days Quadrilaterals have four sides $1 \text{ cm} = 10 \text{ mm}$
Mathematical rules or formulae	Volume = length x width x height

As discussed earlier in this chapter, being fluent in these maths facts really does help to make learning maths easier. Without knowing the rules for calculating volume or finding the mean, we won't get very far. Equally, solving problems involving measure often requires knowing equivalences between different units of measure. As such, many of these mathematical facts form the prior knowledge that children need to be secure with in order to access many other areas of maths.

As a Year 6 teacher, I certainly recognised the importance of my class being fluent with these maths facts. Consequently, building fluency with mathematical statements and rules alongside building fluency with number facts became a daily priority in my classroom. Knowing that children will be taking end-of-key-stage assessments certainly brings a focus upon maths facts to the fore. However, ultimately, it is not how pupils do in these assessments that makes a difference, but how we are preparing children for the next stage in their learning. As such, the focus on retrieval and retention of knowledge to build fluency is highly relevant to all teachers.

Knowledge of number facts

As we established earlier, number facts are the essential building blocks needed to build fluency with so many other areas of maths. So, what number facts do pupils need to be secure with?

In my first few years of teaching, I used the National Curriculum to help me to work out what number facts to teach my Year 1 class. For Year 1, with addition and subtraction, the National Curriculum states that pupils should be able to 'represent and use number bonds and related subtraction facts within 20' (DfE, 2021).

Reading this, I diligently took to teaching my Year 1s their number bonds to 10 and then how to apply these facts to recall number bonds to 20 and their corresponding subtraction facts. Crucially, I had clearly missed the importance of the words 'within 20', as will become clear. Ensuring that children had the opportunity to practise recall of their number bonds to 10 and 20 at the start of each lesson, I was pleased with the progress that my class were making. Consequently, in the summer term, we moved on to looking at how to add two numbers mentally by bridging through 10. Aware of the importance of developing conceptual understanding of this process, I built up this knowledge gradually, firstly using a tens frame to demonstrate the process of how to add by bridging 10, as in the example in Figure 5.6.

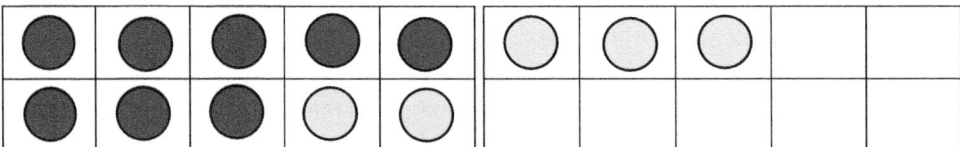

$$8 + 2 = 10$$

$$10 + 3 = 13$$

FIGURE 5.6 A tens frame demonstrating bridging through 10

In order to move pupils towards developing fluency in the abstract, we then used number lines for pupils to record their bridging through 10.

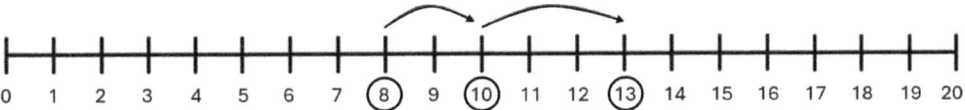

FIGURE 5.7 A number line demonstrating bridging through 10

However, when it came to removing these manipulatives and representations, I found that most of my class struggled to bridge 10 mentally. Reflecting upon why this was, I came to understand the crucial role that number composition plays in building addition and subtraction fluency. For example, when solving 8 + 5, I found that pupils could easily tell me that we needed 2 more to make 10. This was because we had spent most of the year drilling number bonds to 10. However, what they struggled with was working out how many more of the 5 they still had left to add. This was because they did not have a secure understanding of number composition. They didn't securely know that 5 is made up of 2 and 3. It was from this that I learned that a secure knowledge of number composition within 10 – and not just number bonds to 10 – is crucial in developing addition and subtraction fluency. This has since been reflected in the Early Years curriculum, where there is now a much deeper focus upon developing knowledge of number composition within 10 (DfE, 2024).

Progression in number facts

While the National Curriculum tells us the broad brush strokes of what pupils need to know in terms of number facts, it does not break these down into their constituent parts. Further, number fact knowledge is scattered throughout the National Curriculum, with some knowledge appearing within place value, some under addition and subtraction and some under multiplication and division. It was because of this that I decided to write a document of progression in number facts for my school. I wanted a document that would clearly state, all in one place, what facts children need to know and that would break this knowledge down into its key parts.

The breakdown and sequencing of addition and subtraction facts within 20 is taken from Clare Christie's work on building number sense with addition and subtraction facts (2017), where she has taken all of the facts within 20 and split them into groups as shown in Figure 5.8.

Adding 1	Bonds to 10	Adding 10	Bridging/ compensating
Adding 2	Adding 0	Doubles	Near doubles

+	0	1	2	3	4	5	6	7	8	9	10
0	0+0	0+1	0+2	0+3	0+4	0+5	0+6	0+7	0+8	0+9	0+10
1	1+0	1+1	1+2	1+3	1+4	1+5	1+6	1+7	1+8	1+9	1+10
2	2+0	2+1	2+2	2+3	2+4	2+5	2+6	2+7	2+8	2+9	2+10
3	3+0	3+1	3+2	3+3	3+4	3+5	3+6	3+7	3+8	3+9	3+10
4	4+0	4+1	4+2	4+3	4+4	4+5	4+6	4+7	4+8	4+9	4+10
5	5+0	5+1	5+2	5+3	5+4	5+5	5+6	5+7	5+8	5+9	5+10
6	6+0	6+1	6+2	6+3	6+4	6+5	6+6	6+7	6+8	6+9	6+10
7	7+0	7+1	7+2	7+3	7+4	7+5	7+6	7+7	7+8	7+9	7+10
8	8+0	8+1	8+2	8+3	8+4	8+5	8+6	8+7	8+8	8+9	8+10
9	9+0	9+1	9+2	9+3	9+4	9+5	9+6	9+7	9+8	9+9	9+10
10	10+0	10+1	10+2	10+3	10+4	10+5	10+6	10+7	10+8	10+9	10+10

FIGURE 5.8 Addition and subtraction facts (based on Christie, 2017)

This provides a useful overview of which groups of facts within 20 can naturally be taught together. For instance, by looking at the 'Adding 1' facts together, pupils learn what it means to find 1 more than a number, which in turn helps to build retention of this group of facts.

TABLE 5.2 Progression in number facts (based on Christie, 2017* and DfE, 2021)

Reception
• Counting forwards and backwards in 1s to 20
• Subitise (recognise quantities without counting) up to 5 • Automatically recall number bonds up to 5 (including subtraction facts) • Number bonds to 10
• Doubling facts (up to 5 + 5)
Year 1
• Counting to and across 100, forwards and backwards, beginning with 0 or 1 or from any given number • Counting forwards and backwards in even and odd numbers (to 20)

TABLE 5.2 (Continued)

Addition and subtraction (within 10)*	Alongside*
1 Adding 1 (e.g. 7 + 1 *and* 1 + 7) 2 Doubles of numbers to 5 (e.g. 4 + 4) 3 Adding 2 (e.g. 4 + 2 *and* 2 + 4) 4 Number bonds to 10 (e.g. 8 + 2 *and* 2 + 8) 5 Adding 10 to a number (e.g. 5 + 10 *and* 10 + 5) 6 Adding 0 to a number (e.g. 3 + 0 *and* 0 + 3) 7 Near doubles (e.g. 3 + 4 *and* 4 + 3) 8 The ones without a family! (5 + 3, 3 + 5, 6 + 3, 3 + 6)	Partitioning 2, 3, 4, 5, 6 and 10 Partitioning 7, 8 and 9 Linking addition facts to corresponding subtraction facts

- Counting in multiples of 2 (to 20)
- Counting in multiples of 5 (to 50)
- Counting in multiples of 10 (to 100)

Year 2

- Counting in multiples of 2, 5 and 10
- Counting on in 10s from any number, forwards and backwards
- Counting forwards and backwards in odd and even numbers (to 100)

Addition and subtraction (bridging 10)*	Alongside*
1 Doubles of numbers to 10 (e.g. 7 + 7) 2 Near doubles (e.g. 5 + 6 *and* 6 + 5) 3 Bridging (e.g. 8 + 4 *and* 4 + 8) 4 Compensating - Recall and use addition and subtraction facts to 20 fluently - Derive and use related facts up to 100	Partitioning 11 to 20 into single-digit addends (e.g. 9 + 2 = 11) Linking addition facts to corresponding subtraction facts

Multiplication facts	Alongside
1 Recall 10 times table facts 2 Recall 5 times table facts 3 Recall 2 times table facts Making connections with other tables facts (e.g. 10 x, 5 x)	Linking multiplication facts to corresponding division facts

Year 3

- Count in multiples of 4, 8, 50 and 100
- Count up and down in tenths

Multiplication facts	Alongside
1 Recall 4 times table facts 2 Recall 8 times table facts 3 Recall 3 times table facts Making connections with other tables facts (e.g. 2 x, 4 x, 8 x)	Linking multiplication facts to corresponding division facts Using commutativity and associativity to derive related facts (e.g. $30 \times 2 = 60$, $60 \div 3 = 20$ and $20 = 60 \div 3$)

(Continued)

TABLE 5.2 (Continued)

Year 4
<ul><li>Count in multiples of 6, 7, 9, 25 and 1,000</li><li>Count backwards through zero to include negative numbers</li><li>Count up and down in tenths and hundredths</li></ul>

Multiplication facts (up to 12 x 12)	Alongside
1 Recall 6 times table facts 2 Recall 7 times table facts 3 Recall 9 times table facts 4 Recall 11 times table facts 5 Recall 12 times table facts Making connections with other tables facts (e.g. 3 x, 6 x, 9 x)	Linking multiplication facts to corresponding division facts Using commutativity and associativity to derive related facts (e.g. $30 \times 2 = 60$, $60 \div 3 = 20$ and $20 = 60 \div 3$)

Use place value, known and derived facts to multiply and divide mentally, including: <ul><li>multiplying by 0 and 1</li><li>dividing by 1</li><li>multiplying together three numbers</li><li>recognising and using factor pairs and commutativity in mental calculations</li></ul>

Year 5
<ul><li>Count forwards or backwards in steps of powers of 10 for any given number up to 1,000,000</li><li>Count forwards and backwards with positive and negative whole numbers, including through zero</li><li>Continue to practise counting forwards and backwards in simple fractions</li></ul>

<ul><li>Continue to use all the multiplication tables to calculate mathematical statements in order to maintain fluency</li><li>Identify multiples and factors, including finding all factor pairs of a number and common factors of two numbers</li><li>Recall prime numbers up to 19</li><li>Multiply and divide numbers mentally, drawing upon known facts</li><li>Multiply and divide whole numbers and those involving decimals by 10, 100 and 1,000</li><li>Recognise and use square numbers and cube numbers, and the notation for squared (^{2}) and cubed (^{3})</li></ul>

Year 6
<ul><li>Count forwards and backwards in multiples with positive and negative values, including through zero</li><li>Counting in time/g/p/m/cm, etc. (bridging to cm)</li></ul>

<ul><li>Continue to use all the multiplication tables to calculate mathematical statements in order to maintain fluency</li><li>Perform mental calculations, including with mixed operations and large numbers</li><li>Identify common factors, common multiples and prime numbers</li><li>Use knowledge of the order of operations to carry out calculations involving the four operations</li><li>Multiply and divide numbers by 10, 100 and 1,000, giving answers up to three decimal places</li></ul>

This progression in number facts can either be used by an individual class teacher to help them focus in upon the key number facts that need to be taught within their year group, or shared across the school. I found it particularly helpful to share this with my own staff team to help them to identify what key facts needed to be taught and to ensure that there was clear progression in the teaching of number facts across the school. When sharing this with staff, it is important for teachers to consider where their pupils are currently, rather than simply teaching the facts listed for their year group. One way of doing this is to get your staff team to look back at the previous year group(s) and identify what their pupils are already fluent with and where any gaps might be. The priority should then be teaching and securing number fact knowledge from previous year groups, before moving on to the facts within their own year group.

Once teachers are clear as to which facts should be taught, the focus should then be on prioritising the building of factual fluency. As discussed at the start of this chapter, building factual fluency is vital to ensure that children are able to access and have success with so many other areas of the mathematics curriculum. As such, we need to ensure that regular practice of these facts is a priority. As a teacher, it took me a while to realise this. Often, during busier times of the school year, I would end up finding that I didn't have as much time for maths as normal. Perhaps we would need to fit in a nativity rehearsal or assembly would run on. In this situation, it would be all too easy to think, 'I don't have time for number fluency. I'll just skip that part for today.' I think that this can often end up being the case. Teachers can end up prioritising their maths lesson so that there is 'evidence of learning' in books, and not prioritising their maths fluency session – not prioritising the element that is likely to make the most difference to our pupils. Fluency has often been described as 'phonics for maths' and I strongly believe that it should be viewed in this way. If there is 'not enough time', maths fluency is the thing that needs to take priority. Here are some recommendations that I have found effective in planning when number facts should be taught:

1. **Timetable sessions in advance:** If they are planned in, they are more likely to happen!

2. **Plan what to teach when:** Use the fluency progression guidance as a starting point and map out when in the year you intend to teach each group of facts.

3. **Little and often is more effective than once a week:** I started out by having a lesson a week for fluency, but ultimately found that this was not as effective as ten to 15 minutes each day.

4. **Focus on introducing one group of facts at a time:** To reduce cognitive load, it can be helpful to gradually introduce new facts. Focusing upon one group of facts or one times table at a time can be a helpful way in which to do this.

5. **Build in time for consolidation:** Don't assume that once something has been taught, it will be remembered. Focus upon building security with recall of a group of facts, but remember to go back and revisit the facts that pupils have already learned.

How should children learn number facts?

So, how should these facts be taught? As suggested earlier in the chapter, teaching facts in a way that helps to embed conceptual understanding is important. For instance, rather than rote-learning all number facts within 20 in isolation, it is more effective to teach groups of facts together so that pupils can make connections between different facts and understand the underlying rules and patterns. This can be done using representations and structures to draw attention to rules and patterns. For instance, when learning the 'Adding 1' facts, pupils could be encouraged to represent each fact in turn on a tens frame or number line to physically show what it means to add 1 more to a number.

It is also worth considering the prerequisite knowledge that pupils need in order to gain fluency with each group of facts. For instance, adding 2 to a number involves being able to identify the next multiple of 2. This can be supported by regularly getting pupils to count in odd and even numbers. By looking at patterns between facts, pupils can then begin to identify that when adding 2 to an even number, they will always get the next multiple of 2.

As mentioned earlier, as well as learning groups of addition facts, it is also important for pupils to build their knowledge of number composition – for instance, being able to identify the different ways of making 7. It is useful that there is now a much stronger focus on building number composition within the Early Years. The starting point should be building children's understanding of number composition within 5. This can be done in different ways. Recently, there has been a big focus on getting children in the Early Years to subitise numbers. This means being able to recognise how many there are without the need to count in ones. This can be done by using different arrangements of dots and getting children to say how many they can see. When encouraging children to subitise, it is important to draw children's attention away from simply counting the set. One way in which to do this is by putting an image up for one or two seconds so that pupils don't have time to count how many there are individually. Once children have said how many dots they can see, return to the dot pattern and ask children how they knew that there were that many. This helps children to focus on the number composition. Consider the example in Figure 5.9.

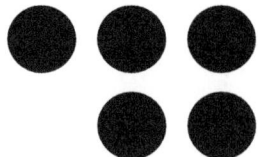

FIGURE 5.9 Dot patterns can develop subitising and number composition

A child might say, 'I know there are 5 because I can see 4 and 1 more.' Or, 'I know there are 5 because I can see 3 on the top row and 2 underneath.' Or, 'I know there are 5 because I can see one less than 6.'

Developing the flexibility to see 5 in different ways helps to strengthen children's knowledge of number composition by helping them to identify what 5 is made up of. Additionally, by providing opportunities for pupils to articulate their thinking, we are helping children to begin to reason mathematically. This can also be useful for those who couldn't subitise 5 straight away. Hearing other children describe what they saw helps them in turn, as they begin to build their knowledge of different ways of seeing 5.

It is also important to be aware of which facts pupils should be able to recall automatically and which facts may require quick strategies to recall. While recall of all of the Year 1 facts should be automatic, many of the Year 2 facts are facts where pupils may need to use a quick strategy to calculate. Consider how you would solve '8 + 9' for instance. You may know it automatically or you may have used a quick strategy to calculate it. On her website Number Sense Maths, Clare Christie discusses three key strategies that pupils may use to add numbers that cross 10: bridging 10, near doubles and compensating (Number Sense Maths, 2024).

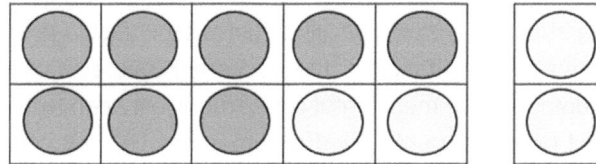 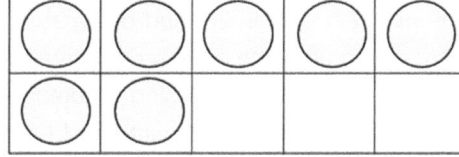

$$8 + 9 = 17$$

FIGURE 5.10 A tens frame used to bridge 10

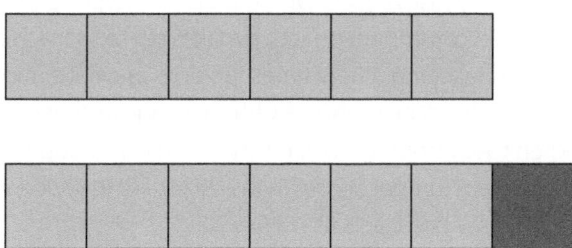

$$6 + 6 = 12$$

$$So, 6 + 7 = 13$$

FIGURE 5.11 Bar models used to show near doubles

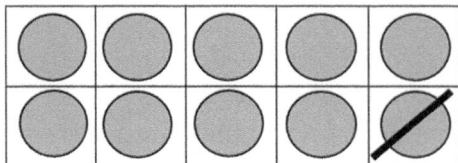 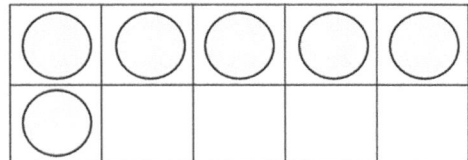

10 + 6 = 16
So, 9 + 6 = 15

FIGURE 5.12 Tens frames showing a compensation strategy

Although these approaches should be taught separately initially, it is important for children to get to a stage where they can use a combination of methods to add numbers that cross 10. For example, all three of these methods could be successfully used to calculate 8 + 9. It is also important for pupils to recognise when a particular method is appropriate – for instance, that a near doubles method can be used to add two consecutive numbers.

How should children learn times table facts?

The prospect of learning and being able to accurately recall all times table facts up to 12 x 12 can often be quite a daunting one – 144 facts is a lot to learn! However, it can be argued that by encouraging children to make meaningful connections between facts, we can cut down on the number of facts that pupils need to remember. For instance, if pupils connect their knowledge of 6 x 7 = 42 with 7 x 6 = 42, then they have fewer facts to remember. These are essentially the same times table, so pupils do not need to learn these facts independently of each other. And given that this is the case for all of our times tables except for square numbers, where the same factor appears twice, then our list of 144 facts is actually just 78 facts with the factors reversed. To ensure that pupils make connections between commutative facts, it makes sense to teach commutative facts together or at least, when teaching a new times table, to draw attention to the facts that pupils have already encountered before. Clare Christie suggests that one way of doing this is to encourage children to say both sets of facts the same way around, so they learn to link commutative facts together (Number Sense Maths, 2024). For instance, whether pupils see '4 x 5' or '5 x 4', they would say both facts as '4 x 5 is 20'.

Further, although our National Curriculum states that pupils need to be able to know and recall all times tables up to 12 x 12, some of these are more useful than others. It is the facts up to 9 x 9 that pupils need to know in order to mentally calculate all of the other times table facts. Considering both the commutative nature of times table facts and the facts that pupils need to be most fluent with, we can narrow down the number of facts that pupils need to learn from 144 to just 36. These are identified within the 'ready-to-progress' criteria guidance (DfE, 2020).

2 x 2							
2 x 3	3 x 3						
2 x 4	3 x 4	4 x 4					
2 x 5	3 x 5	4 x 5	5 x 5				
2 x 6	3 x 6	4 x 6	5 x 6	6 x 6			
2 x 7	3 x 7	4 x 7	5 x 7	6 x 7	7 x 7		
2 x 8	3 x 8	4 x 8	5 x 8	6 x 8	7 x 8	8 x 8	
2 x 9	3 x 9	4 x 9	5 x 9	6 x 9	7 x 9	8 x 9	9 x 9

FIGURE 5.13 The 36 times table facts that pupils need to learn (based on DfE, 2020, p. 332)

Focusing the recall of times table facts on these 36 key facts helps to reduce cognitive load by reducing the number of individual facts that pupils need to learn and remember. Additionally, saying commutative facts in the same way helps pupils to make links between these facts (Number Sense Maths, 2024). Another way of reducing the amount that pupils need to remember is by encouraging them to just say the factors and products for each times table. For instance, instead of saying '5 times 9 equals 45', pupils would just say '5, 9, 45'. This was an approach that I saw used by pupils in Shanghai and is also used within the NCETM's 'Mastering number at KS2' programme (NCETM, 2024a). Repetition of the factors and product alone is useful because there is a smaller chunk of information for pupils to retain. Additionally, it helps pupils to develop strong links between the factors and product of each times table fact, which in turn aids recall of these facts.

In a similar way to learning addition facts, a 'little and often' approach works well – for instance, asking pupils to complete a set of times table questions daily and then getting pupils to chant the facts as you mark them together. A whole-class approach can be used for this initially. However, once most pupils can confidently recall the majority of their facts, it can then be useful for individuals to begin to focus on practising the times tables that they are still finding tricky until they are fluent with them all.

Knowledge of mathematical statements and formulae

Number facts form one type of factual knowledge pupils need to be fluent with but there are other types of maths facts that can also help to give pupils access to the wider curriculum. These include mathematical statements and mathematical formulae.

In terms of embedding knowledge of mathematical statements and rules, there are a few principles that I have found particularly useful. Here are some of my top tips:

1. **Embed in context:** Maths facts should not be introduced in isolation, but taught in a meaningful way that helps to build understanding.

2. **Link to prior learning:** Prior learning is more meaningful when linked to relevant new learning. In addition to this, connecting new learning to prior learning helps learners to make sense of this new knowledge.

3. **Repeat within the lesson:** Identify the key facts that pupils need to know at the start of the lesson and ensure that there are regular opportunities to reinforce these and check recall throughout the lesson.

4. **Repeat across lessons:** Plan in regular opportunities to revisit and review these key maths facts until they can be recalled automatically.

5. **Make effective use of formative assessment:** When consolidating recall of facts, ensure that there are ways to check that the understanding of all pupils is secure, and intervene when it is not.

To combat the effects of Ebbinghaus's forgetting curve, spaced retrieval is key. Of course I'm going to revisit equivalences between units of measure at the start of a lesson focused around solving problems involving different units of measure. This allows for learning to be embedded in context and gives ample opportunities for pupils to apply their knowledge of these maths facts to help them to solve problems. However, I can't then wait another year until I teach measures again to consolidate this knowledge. Learning should be consolidated throughout a topic, but then revisited later in the year. This is particularly the case if you are following a block-based curriculum structure. Teaching topics in blocks allows time for building a real depth of understanding. But the principles of spaced practice emphasise the importance of revisiting this knowledge regularly.

Having a list of what these key maths facts are provides a useful starting point. Table 5.3 provides an example of the key maths facts that I put together for my Year 6 class. It is certainly not a complete list of all the maths facts that pupils need to know by the end of Key Stage 2. However, it comprises the facts that I identified as being particularly important, both in terms of preparation for SATs and, more crucially, for ensuring that pupils were ready for secondary school. These were the facts that I saw as key for providing access to the wider mathematics curriculum.

TABLE 5.3 Key maths facts

Measures	Months	Shape properties
1 cm = 10 mm 1 m = 100 cm 1 km = 1,000 m 1 kg = 1,000 g 1 litre = 1,000 ml £1 = 100p	**31 days:** January, March, May, July, August, October, December **30 days:** September April, June, November **28/29 days:** February	**Shapes:** Quadrilateral – 4 sides Pentagon – 5 sides Hexagon – 6 sides Octagon – 8 sides **Triangles:**
Time	**FPD equivalences**	Equilateral – all sides/angles the same Isosceles – 2 sides/angles the same Scalene – all sides/angles different **3D shapes:** Vertices – corners Edges – ridges/sides
1 century = 100 years 1 decade = 10 years 1 year = 365 days 1 year = 12 months 1 week = 7 days 1 day = 24 hours 1 hour = 60 minutes 1 minute = 60 seconds	$\frac{1}{10} = 10\% = 0.1$ $\frac{1}{100} = 1\% = 0.01$ $\frac{1}{2} = 50\% = 0.5$ $\frac{1}{4} = 25\% = 0.25$ $\frac{1}{5} = 20\% = 0.2$	
Calculating the mean	**Angles**	**Perimeter, area and volume**
*Find the total: Add together all of the values *Find the average: Divide the total by the number of values	Right angle = 90° Triangle = 180° Quadrilateral = 360° Straight line = 180° Full circle = 360°	**Perimeter of a rectangle** = 2 x (length + width) **Area of a rectangle** = length x width **Volume of a cube or cuboid** = length x width x height

Having a list of key maths facts provided me with a starting point for identifying the knowledge for which I wanted to prioritise retrieval and consolidation. Once taught in context, I would then plan quick retrieval tasks to use at the start of lessons to reinforce knowledge and recall of these facts. The next section provides some examples that you might like to use with your own class.

Maths facts: Retrieval tasks

Name the type of angle and the number of degrees in each.

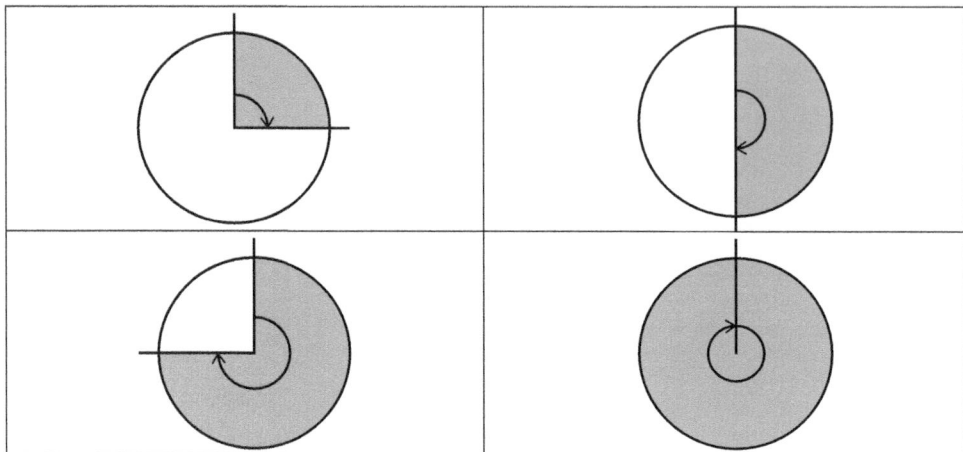

FIGURE 5.14 Demonstrating knowledge of angles

TABLE 5.4 Demonstrating equivalent measures

What are the equivalences for the following units of measure?	
1 km = __ m	1 year = __ days
1 m = __ cm	1 year = __ months
1 cm = __ mm	1 fortnight = __ days
1 kg = __ g	1 day = __ hours
1 litre = __ ml	1 hour = __ minutes
£1 = __ p	1 minute = __ seconds

TABLE 5.5 Demonstrating equivalent fractions, decimals and percentages

Complete the table, showing the decimal and percentage equivalences.		
Fraction	**Decimal**	**Percentage**
$\frac{1}{2}$		
$\frac{1}{4}$		

Table 5.5 (Continued)

$\dfrac{1}{5}$		
$\dfrac{1}{10}$		
$\dfrac{1}{100}$		

This task can be easily adapted, so that pupils are given either the fractions, decimals or percentages and asked to recall the missing equivalences for the remaining two categories. Alternatively, try using a mixed grid like Table 5.6.

TABLE 5.6 Demonstrating equivalent fractions, decimals and percentages

Fraction	Decimal	Percentage
	0.5	
$\dfrac{1}{4}$		
		20%
	0.1	
		1%

Once children are confident with recalling these root facts, extending their practice to derive related equivalences can be useful.

TABLE 5.7 Practising decimal fraction equivalents

Calculate the decimal equivalences of the following fractions:		
$\dfrac{1}{2} =$	$\dfrac{2}{2} =$	$\dfrac{3}{2} =$
$\dfrac{1}{4} =$	$\dfrac{2}{4} =$	$\dfrac{3}{4} =$
$\dfrac{1}{5} =$	$\dfrac{2}{5} =$	$\dfrac{4}{5} =$
$\dfrac{1}{10} =$	$\dfrac{2}{10} =$	$\dfrac{6}{10} =$
$\dfrac{1}{100} =$	$\dfrac{2}{100} =$	$\dfrac{6}{100} =$

Notice that, at this stage, questions are carefully scaffolded so that the root fact needed in each case is identified. Use of variation means that children can see the connection between calculating the decimal equivalences of $\frac{1}{5}$ and $\frac{2}{5}$ and then $\frac{4}{5}$. As pupils develop fluency with this, we can then remove the scaffolding provided by the variation to see whether pupils can apply what they have learned. For instance, in the retrieval activity in Table 5.8, pupils have to independently identify which unit fraction might help them to find the equivalences of the non-unit fractions.

TABLE 5.8 Making connections through retrieval practice

Calculate the decimal equivalences of the following fractions:		
$\frac{2}{2} =$	$\frac{4}{2} =$	$\frac{8}{2} =$
$\frac{2}{4} =$	$\frac{4}{4} =$	$\frac{3}{4} =$
$\frac{2}{5} =$	$\frac{4}{5} =$	$\frac{6}{5} =$
$\frac{2}{10} =$	$\frac{4}{10} =$	$\frac{9}{10} =$
$\frac{2}{100} =$	$\frac{4}{100} =$	$\frac{99}{100} =$

As pupils get older, there is also value in them taking greater responsibility for their own learning. One way I found in which to encourage this was by getting my class to make their own 'maths fact checkers'. These were normally grouped thematically so that similar facts could be practised together. To create these for your class, simply fold an A4 piece of paper in half to create two long, thin rectangles. The question or prompt e.g. '1m = __ cm' (see Figure 5.15 for ideas) can then be written on the top with the answer e.g. '100 cm' written underneath. Pupils can then fold them in half so that the questions are on the top and the answers are underneath. Finally, they will need to cut along the black lines until they reach the fold, ensuring that they only cut the top sheet and not the sheet with the answers underneath. Pupils can then use these either with a partner, to test one another's recall or individually, lifting the flap to check whether each answer is correct as they go. I used to find these particularly useful for homework practice, where pupils knew that they would be tested on recall of these facts the following week. Additionally, they are great for keeping in trays so that pupils have them to hand if you have a spare few minutes for retrieval practice at the beginning or end of a lesson.

1 km = _m	1000 m	Hexagon	6 sides
1 m = _ cm	100 cm	Pentagon	5 sides
1 cm = _mm	10 mm	Octagon	8 sides
1 kg = _ g	1000 g	Quadrilateral	4 sides
1 litre = _ ml	1000 ml	Equilateral triangle	All sides / angles same
1 year = _ days	365/366 days	Isosceles triangle	2 sides / angles same
1 day = _ hours	24 hours	Scalene triangle	All sides / angles different
1 hour = _ mins	60 minutes	Degrees in triangle	180°
1 minute = _ s	60 seconds	Degrees in quadrilateral	360°
30 days in…	Sept, Apr, June, Nov	Straight line	180°

FIGURE 5.15 Maths fact checker templates

Alongside this, I used to share the whole list of maths facts with my Year 6 class after all of the facts had been introduced and taught in context. I would then get them to individually identify the facts that they already knew and the facts of which they were still unsure. A good way in which to do this is to get pupils to work in pairs to assess each other's knowledge and recall, e.g. 'How many degrees in a right angle?' Then simply tick or cross which facts are known and which are not. This then gives pupils a set of focus facts to practise. In a similar way to the previous activity outlined, pupils can then use this to create their own personally tailored knowledge checkers. Using a blank template, they simply write one part of the fact on the front and its equivalence or definition underneath. This is a really useful way in which to target the facts pupils are still not secure with, so that they can focus upon consolidating recall of the facts that are most relevant to them.

As mentioned earlier, this is not a full list of facts that pupils need to know. While it includes many of the most important facts, it largely focuses upon those involving equivalences or definitions. Some facts not identified within this list are those for which having a visual image to refer to is necessary. These might include facts such as identifying the names and properties of shapes or describing the meanings of key terminology. The following figures provide some examples.

Name the following quadrilaterals and identify their properties:

FIGURE 5.16 Retrieval of the properties of quadrilaterals

Name the following 3D shapes and record how many faces and vertices there are in each one:

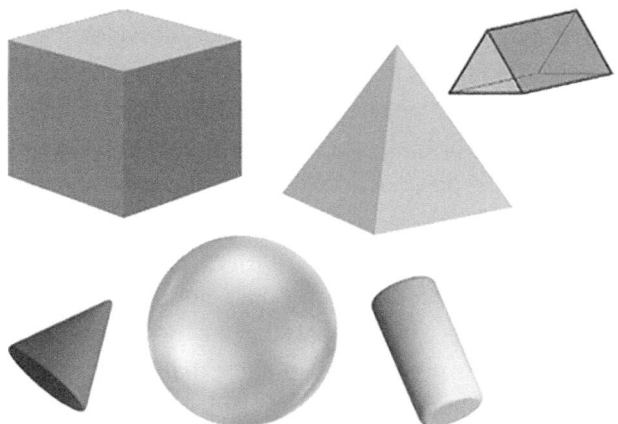

FIGURE 5.17 Retrieval of the properties of 3D shapes

Name the type of triangle and identify its properties:

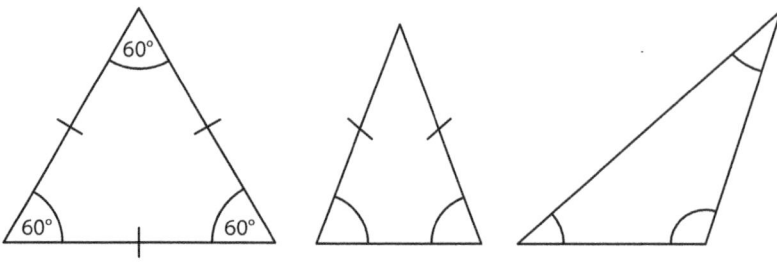

FIGURE 5.18 Retrieval of the properties of triangles

What type of angle is shown in the images below?

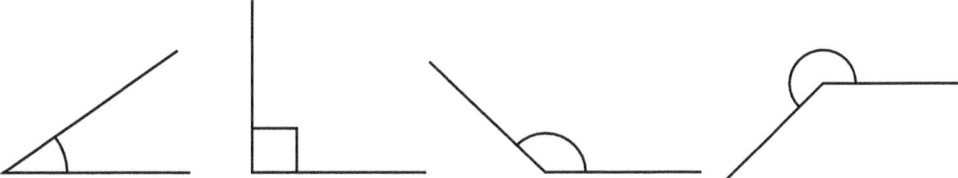

FIGURE 5.19 Retrieval of the types of angles

Knowledge of facts, methods and strategies: What to prioritise?

Over recent years, there has been an increased emphasis upon the use of retrieval activities within the classroom. However, as teachers are often busy and time-poor, retrieval is not always carried out in a very systematic way. For instance, there are many commercially available schemes and products that have sets of retrieval activities that you can download and use straight away. While this is certainly a time-saver, do all of these materials prioritise consolidation of the maths facts and methods that are most important for your pupils? For instance, I might regularly ask my class to practise reading Roman numerals, but is this the knowledge that I think is going to equip them well for starting secondary school? Absolutely not!

Another difficulty that teachers often face is the sheer amount of content that there is to cover within the primary mathematics curriculum. If our aim is for pupils to gain mastery of this curriculum, it seems a near impossible task when there is simply so much to cover! Further, this problem can often lead teachers to do the opposite of what we are advocating within this book. When there is so much content to 'get through', it can be all too easy for us to feel that we do not have time to stop – thinking perhaps, 'I still have another nine steps to cover in this unit and I can't afford to spend another lesson on this,' or 'Most children seem to have got it and there is so much still to do, so I should probably just move on.' While well-intentioned, this can lead to having too high a focus upon task completion rather than prioritising the embedding of deep and secure understanding.

So how do we ensure that there is time to focus on the things that really matter and to ensure that this knowledge is deeply embedded? Which knowledge should be prioritised and how do we determine this?

As mentioned earlier, knowledge of key maths facts, including number facts, provides essential building blocks, forming the foundations for proficiency with so many areas of maths. Another useful tool that we have to hand to help us to identify the most important knowledge is the Department for Education's 'ready-to-progress' criteria guidance (DfE,

2020). This guidance aims to summarise the most important parts of the mathematics curriculum. It highlights the parts of the curriculum that are absolutely essential for pupils to be fluent with in order to be 'ready to progress' to the next year group. As a teacher, I found it really helpful when this guidance was first introduced. Although it was introduced during Covid and provided a useful tool for addressing gaps in knowledge, it was never intended to be a Covid recovery document. Its purpose goes beyond this. Yes, it can certainly be used to track back and identify areas of the curriculum with which pupils are not yet secure. But, ultimately, it works best if these areas are the areas in which we are encouraging teachers in every year group to prioritise fluent and secure understanding in the first place.

Each statement summarises an end point in which children are expected to be fluent by the end of the year. Alongside this, each objective or end point is accompanied by guidance notes to support with teaching. The NCETM have also produced slideshows to go with each objective (NCETM, n.d.). I have found both the guidance documents and the slideshows useful to support with my own teaching of these objectives and to ensure that these areas are prioritised in my own classroom practice. However, what I would like to focus on is how this guidance can be used to shape our approach towards building fluency with facts and methods.

The first thing to say is that practice with end points does not build fluency with end points. Take criterion 3NPV-3 for instance, a number and place value criterion for Year 3, which states that pupils should be able to 'Reason about the location of any three-digit number within the linear number system' (DfE, 2020, p. 9). Giving pupils lots of practice at marking numbers on marked and unmarked number lines is not the best way of building proficiency with this end point. There are several skills within this larger end point with which pupils need to build proficiency. For instance, building proficiency with this involves:

- knowing how to accurately identify the midpoint on a number line
- knowing how to calculate what each interval on a number line is worth
- knowing how to identify which multiple of 10 or 100 a number is closest to.

Proficiency with the parts is what helps to build proficiency with the whole. As such, our approach to retrieval and recall of knowledge should reflect this. We should begin by securing fluency with each of the individual components, before ensuring that pupils can combine their understanding of these components to reach the intended end point.

Knowledge of methods and strategies

Although we might talk about the importance of fluency with concepts, facts, methods and strategies as four distinct categories, in reality it is often difficult to completely separate these because our ability to use these four types of knowledge is often interlinked. Indeed, to completely separate them feels artificial. As mentioned earlier, it is our knowledge and fluency with facts and methods that helps us to build fluency with strategies.

Consider the following problem:

Ben has two packets of sweets. He eats $\frac{4}{5}$ of a packet. Sasha eats $\frac{3}{5}$ of a packet. How much do they have remaining?

Solving such a problem relies upon pupils having secure knowledge of both key facts (e.g. number facts within 10, equivalence of 1 and $\frac{5}{5}$) and key methods (e.g. converting whole numbers to improper fractions, addition and subtraction of fractions with the same denominator). Table 5.9 shows three different ways in which a child might solve this problem.

TABLE 5.9 Three different strategies to solve the same problem

Strategy 1	Strategy 2	Strategy 3
$\frac{4}{5} + \frac{3}{5} = \frac{7}{5}$	$\frac{4}{5} + \frac{3}{5} = \frac{7}{5}$	$2 - \frac{4}{5} = 1\frac{1}{5}$
$2 = \frac{10}{5}$	$2 - \frac{7}{5} = 2 - \frac{5}{5} - \frac{2}{5}$	$1\frac{1}{5} - \frac{3}{5} = 1\frac{1}{5} - \frac{1}{5} - \frac{2}{5}$
$\frac{10}{5} - \frac{7}{5} = \frac{3}{5}$ of a packet	$= \frac{3}{5}$ of a packet	$= \frac{3}{5}$ of a packet

In order to develop fluency with each of these strategies, pupils first need to understand why each strategy works. Additionally, they need to have secure knowledge of the facts and methods that underpin each strategy and to understand how they can be used in conjunction with one another. Some pupils might be able to use their wider understanding of facts, methods and concepts to arrive at these strategies independently. However, most children will acquire fluency with knowing which strategies to use and when through direct modelling, discussion and practice.

Selecting the most appropriate methods and strategies to model

If modelling methods and discussing strategies is so important for the development of pupils' own fluency, then it follows that teacher subject knowledge is vital in helping our pupils to attain this fluency. In order to select the best strategy, we need to know which methods can be used and to be able to evaluate which methods are the most effective.

To give an example, consider how you would calculate:

$$12 - 5 = 7$$

The following figures demonstrate three possible ways:

1. counting back in 1s

2. bridging through 10

3. subtracting from the 10.

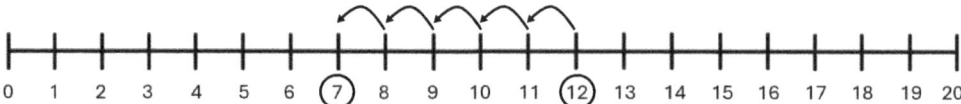

FIGURE 5.20 A number line showing 12 – 5 by counting back in 1s

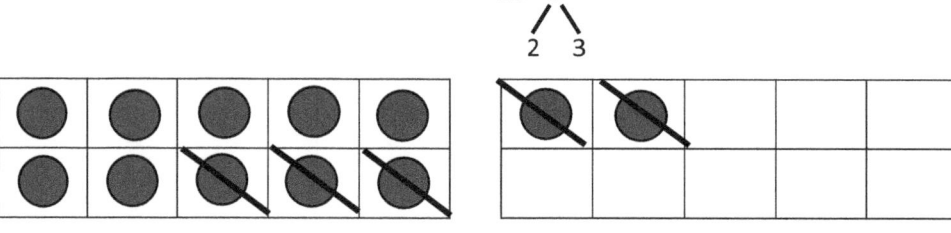

$$12 - 2 = 10$$

$$10 - 3 = 7$$

FIGURE 5.21 Tens frames showing bridging through 10

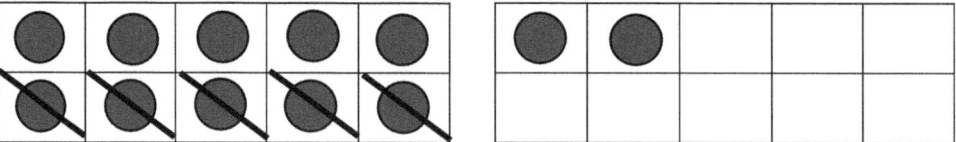

$$10 - 5 = 5$$

$$5 + 2 = 7$$

FIGURE 5.22 Tens frames showing subtracting from the 10

When considering the approaches that we wish to model in the classroom, it is important to consider the following:

- **Efficiency:** Is this an efficient way in which to solve the calculation?
- **Prior knowledge:** Does the pupil have the prior knowledge required to successfully use this method?
- **Flexibility:** Have you discussed and modelled more than one way in which to solve the same problem?

In this case, bridging through 10 and subtracting from 10 are far more efficient than counting back in 1s. Counting back in 1s involves more steps than using partitioning. Further, bridging and subtracting from 10 both mirror the efficient approaches that we would expect adults to use. However, successful use of bridging through 10 and subtracting from 10 requires secure recall of number facts and a secure understanding of number composition. For instance, successfully bridging through 10 in the abstract requires the pupil to have a secure understanding of the composition of 5 (e.g. knowing that 5 can be partitioned into 2 and 3), as well as number bonds to 10 ($10 - 3 = 7$). It is important to recognise that this does not mean that we should be teaching children who do not have the necessary prior knowledge to use inefficient approaches, such as counting back in 1s. In the long run, this approach only leads to a widening gap in attainment, where our most able learners can use quick and efficient methods and struggling learners are forced to use inefficient methods. Rather, the priority must be focusing upon building fluency with the prior knowledge needed to give all children access to the most efficient calculation methods – hence the need to prioritise children's knowledge and recall of key number facts.

The final consideration is concerned with developing flexibility. As mentioned earlier, which strategy is most efficient can often be context-dependent. This is where having more than one method to hand can be an advantage. If pupils only know one method for solving a problem, they will default to using this. Here are some good ways I have found in which to encourage pupils' flexibility with strategies:

- **Method fluency:** Build pupils' proficiency with a range of calculation methods.
- **Modelling different strategies:** Model more than one strategy that can be used to solve the same problem.
- **Retrieval practice:** Once pupils are fluent with a range of methods, provide opportunities for them to apply these to new and unfamiliar contexts, where they have to decide for themselves which strategy they will choose to use.
- **Mathematical discussion:** Foster a language-rich classroom where pupils are regularly encouraged to explain the strategies that they have chosen to use.
- **Evaluation of strategies:** Encourage pupils to evaluate their strategies and consider whether there is a more efficient strategy available.

It is also important to recognise the limitations of our own subject knowledge and how this in turn influences our pupils. For example, if I think about how I would naturally choose to calculate 12 − 5, I would bridge through 10. It was only when I had the opportunity to watch a teacher from Shanghai teaching a Year 2 class that I realised how subtracting from 10 could be used. This was not a method that I had previously been taught and so not a method I would naturally choose to use. This in turn made me reflect upon the limitations of my own subject knowledge. As teachers, it can be easy to default towards modelling the methods and strategies with which we feel most comfortable. However, in so doing, we may well end up limiting the subject knowledge and flexibility of our pupils.

The solution, of course, lies in upskilling our own mathematical subject knowledge. When observing lessons in Shanghai, I was struck by the meticulous nature of the lesson design process. It was clear that teachers invested a lot of time in considering the careful sequencing of knowledge within lessons, as well as taking time to watch one another's lessons and unpick how the mathematics is developed. There were two key messages that I took away from this experience. The first was the importance of the lesson design process. Use of the NCETM's professional development materials is something that I have found incredibly valuable in developing my own subject knowledge around which methods and strategies to model and when (NCETM, 2024b). Secondly, it taught me about the value of watching and analysing maths lessons to learn from others. If pupil outcomes are directly linked to the quality of teaching and learning in the classroom, it makes sense that we take the time to prioritise our own continuing professional development, and developing our own subject knowledge around methods and strategies is an important way in which we can do this.

Key takeaways

We have considered why fluency with number facts is a vital part of building mathematical proficiency. We have established that fluency and conceptual understanding are directly linked, and that fluency involves the ability to work accurately, efficiently and flexibly (Russell, 2007) with facts, methods and strategies. Being intentional about developing our own subject knowledge will help us in turn to provide our pupils with a depth of subject knowledge, so that they can work flexibly to evaluate the efficiency and accuracy of the strategies that they are using within a given context.

Finally, a regular and systematic focus on developing number fact fluency is essential if we are to ensure that all pupils have fair access to the mathematics curriculum in our classrooms.

6 Variation

For me, teaching with variation is the most powerful of all the ideas in this book, but it can be notoriously difficult to develop and embed in classrooms and across schools. In this chapter, I will try to reveal its power in getting children to think mathematically, look for connections and build a rich tapestry of understanding.

Conceptual variation

I'd like you to think of a large grey mammal. It has thick, tough skin. Many of you will be thinking of an elephant at this point, as those are some of the essential features of something being an elephant. Now let's alter your thinking and schema and state that this mammal has tusks but also has flippers rather than feet. Now, perhaps, you're thinking of a walrus. Elephants and walruses both have things in common, but it is not until we know the essential features of what it is and, importantly, of what it is not that we can begin to build the concept of both mammals in our mind.

To learn anything at all, we need to know the essential features of 'it'. But we also need to know about the non-essential features to know what 'it' is not.

I often use the following sequence to introduce this idea of building a concept for the first time. I use an unfamiliar word – cerulean – with adults to really get them thinking, making connections and wondering what the concept may be. I slowly reveal pictures, each of which shows a hue of blue, such as blue hair, blue clothes or blue eyes, and state for each image 'This is cerulean' or 'This is not cerulean'.

As we work through this, adults build an idea that cerulean is a particular shade of blue. Their thinking is confirmed when they see what it is not, such as a red cherry or a green emerald.

This can be an effective way in which to think about introducing new concepts. Now, let's think about the concept of a triangle. We often introduce this to children in the Early Years and Key Stage 1. For many years, I used to show children a triangle that I had quickly found online without much thought, and then proceeded to define what a triangle is. I kept saying that a triangle is a 2D shape with three sides and three vertices. The glaring error that I had made was that I hadn't told children what it is not. Consider the images in Figure 6.1.

FIGURE 6.1 Triangle or not a triangle?

Notice how presenting these images (perhaps one by one to a class of children) allows pupils to consider what a triangle is and is not. This type of conceptual variation opens the discussion into deeper reasoning and allows children to build a concept up over time. It allows us as teachers to guide the thinking of it being a closed shape where all the lines meet; those lines should be straight; and just because a shape has a point, that does not mean that it is a triangle… and so on. We can also draw attention to the non-essential features, such as its size or colour. This type of activity will allow children in Early Years and Key Stage 1 to develop a stronger sense of what a triangle is and is not.

An easy way of implementing this could be with the type of comparison activity in Figure 6.2, with carefully selected representations to draw attention to essential and non-essential features.

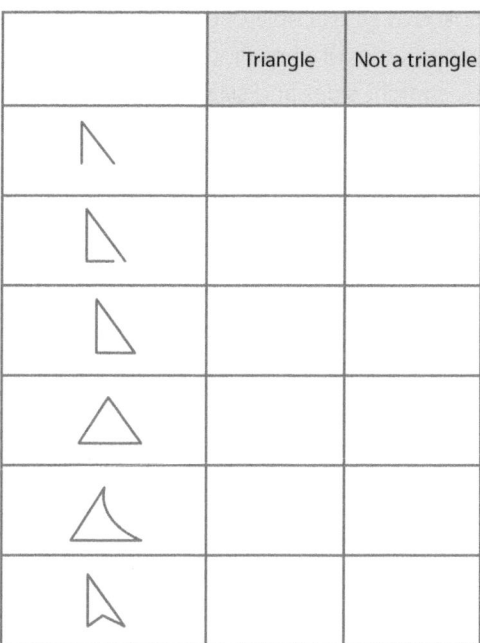

	Triangle	Not a triangle

FIGURE 6.2 A possible activity to check for the essential and non-essential features of a concept

As children progress to Key Stage 2, they will start to look at subcategories of triangles, including right-angled, scalene, isosceles and equilateral. We could repeat a similar activity with children to show the essential and non-essential features of the different types of triangles. Here, we can see how this way of teaching leads to generalisations that children can use again and again.

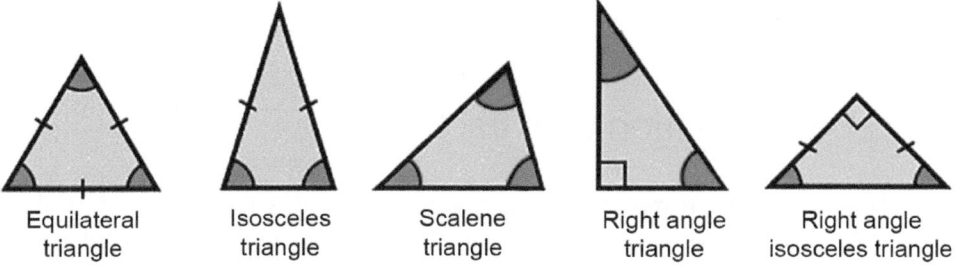

| Equilateral triangle | Isosceles triangle | Scalene triangle | Right angle triangle | Right angle isosceles triangle |

FIGURE 6.3 Types of triangles

I should mention a common point of confusion about variation. Variation is not variety. All lessons will likely have a variety of representations. Variation is about the deliberate changes to the essential and non-essential features of a concept or procedure, to draw attention to mathematical structures.

Now that we have looked at an example of conceptual variation, how would you now teach children about the properties of a hexagon? We might consider the essential features of it being 2D, having six straight sides and six vertices. We might then begin with some familiar shapes.

FIGURE 6.4 Conceptual variation of hexagons – what is the same? What is different?

Here you can see a range of hexagons. The essential features of six sides and six vertices remain, while other non-essential features, such as the length of sides, size, orientation and colour, change.

Conceptual variation with fractions

Let's take a look, in the name of coherence, at fractions across the primary curriculum through the lens of conceptual variation. This can be an area that children (and adults) worry about. In Key Stage 1, we can essentially repeat the process that we considered with hexagons: 'half or not half', 'a third or not a third', 'a quarter or not a quarter', looking for the essential features of each respective fraction as having the relevant amount of equal parts.

We can see how fractions coherently progress from Key Stage 1 to Key Stage 2 and beyond. Planning out examples and non-examples – what a concept is and what a concept is not (a 'non-cept'?) – can be a useful way in which to start implementing conceptual variation into your classroom and school curriculum.

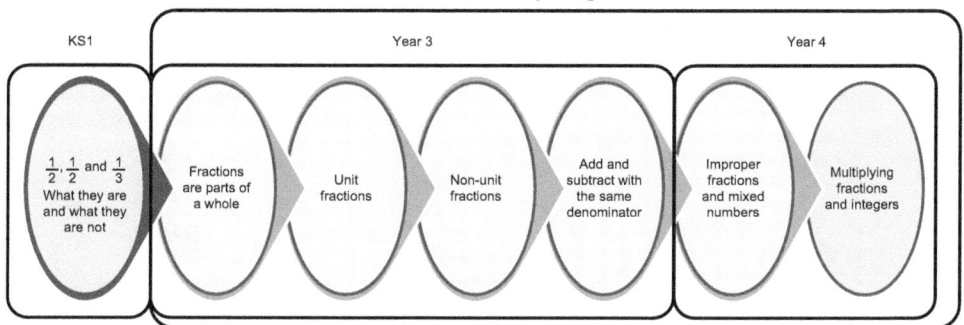

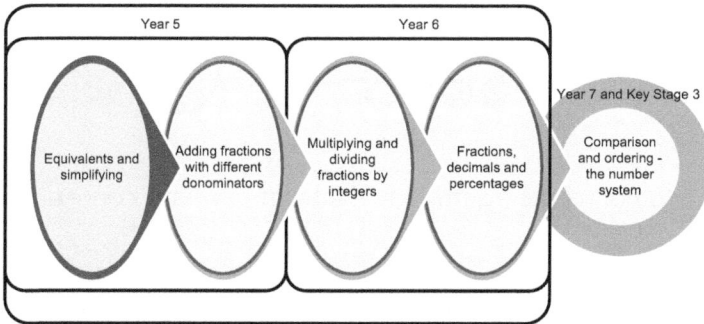

FIGURE 6.5 Coherent progression of fractions

Perhaps an easier way of saying 'essential' and 'non-essential' is 'what is the same?' and 'what is different?' We'll use that from now on in.

In Figure 6.6, you can see how you could show children the concept of what each of the unit fractions is and is not. In doing so, you get to the essential features of each fraction. For $\frac{1}{2}$, we may generalise that there are two equal parts with one part shaded in. The use of well-chosen representations is important to reveal the essential and non-essential features of what you are teaching.

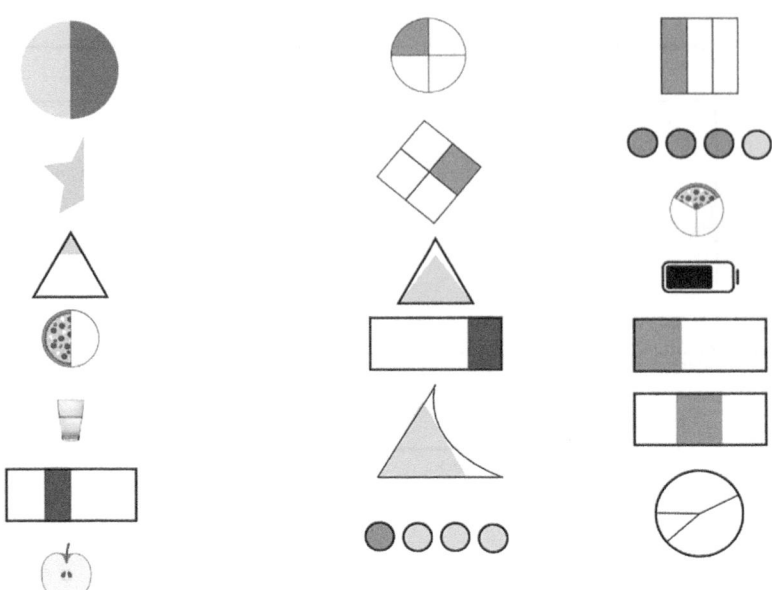

FIGURE 6.6 Concepts and non-cepts – getting to the heart of what a concept is and is not

Let's look at the start of Key Stage 2. Children formally learn that fractions are a part of a whole. Here, we might want to start with non-maths examples. Lots of Year 3 classes learn about continents, countries and the UK, so perhaps this would be useful for exploring the relative sizes of parts of fractions. Consider the images in Figure 6.7. Each image would be shared separately with children, rather than together, and act as a discussion point.

FIGURE 6.7 Building the concept of a fraction being a part of a whole

While giving some examples, such as 'If the world is the whole, Africa is part of the whole', I would also ask for some non-examples, such as 'If the world is the whole, the moon is part of the whole'. Again, we consider what the concept is and what the concept is not, in order to build conceptual understanding. With each image, the whole is getting smaller, as is the relative size of the parts.

As children progress, they start to consider that a whole can be divided into any number of equal parts, and link this directly to fraction notation.

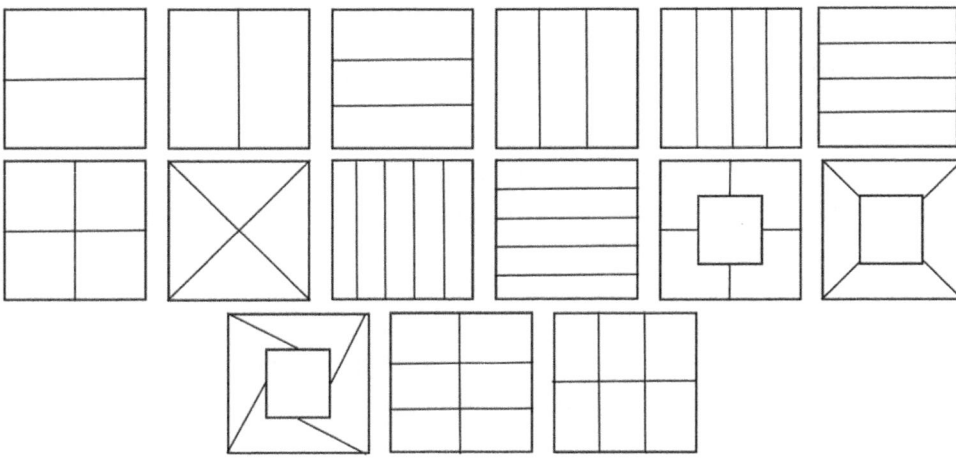

FIGURE 6.8 Look at how this sequence develops with each example – what do we want to draw attention to? What do we want children to notice?

By showing these representations, you are also laying strong foundations for comparing fractions, as children can begin to notice that $\frac{1}{2} > \frac{1}{3}$.

Children may then progress to learn that equal parts do not need to look the same. Consider the example in Figure 6.9, which draws out the generalisation that $\frac{1}{3}$ needs three equal parts.

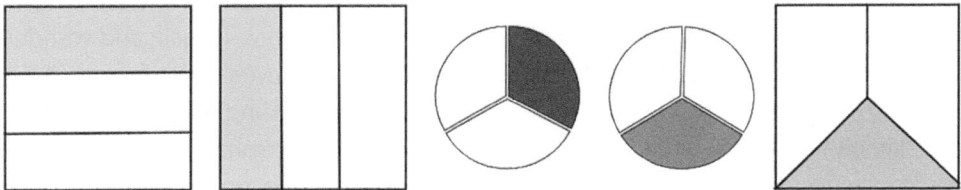

FIGURE 6.9 Same concept but different representations

We may then begin to compare the fractions and generalise, using carefully chosen representations, that the greater the denominator, the smaller the fraction, but we may want to draw out a key misconception around this.

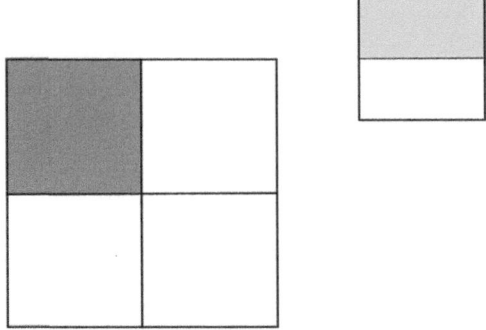

FIGURE 6.10 Comparing fractions and focusing on essential and non-essential features

In Figure 6.10, we are drawing attention to the relative sizes of the whole (just as we did earlier with the non-maths examples) and showing children that $\frac{1}{4} < \frac{1}{2}$. We are teaching that a greater proportion of the smaller representation is shaded than the larger representation. We could ask children to show this with two squares that are the same size, to further strengthen this point. Again, asking 'What is the same? What is different?' can be effective to draw this reasoning out, focusing on the number of equal parts and the number of parts shaded in. Once secure, children can begin to focus on knowing a part to find the whole.

The power of teaching in this way, complementing it with the careful and deliberate use of representations, allows children to think deeply about mathematics and reason accurately. This can unlock mathematics for so many children – and to think that I could have downloaded a ready-made slideshow and told children some definitions of a fraction. Yikes!

I have found, when teaching using variation theory, that I am able to 'think on my feet' in lessons to adapt, personalise, improvise and develop bespoke activities for children within lessons. It is clear how this idea could be developed to allow for a greater depth of understanding. The image in Figure 6.11 shows the same structure but in different contexts, to secure greater depth of understanding at primary school. I used to give all children the first example to work on as a task to reason, draw, explain and wonder. It would be accompanied with the following sentence stem to reveal the structure: 'There are _____ equal parts and _____ part is shaded.' I'd helicopter around the room, checking the climate and temperature of the class. I'd then adapt, personalise, improvise and develop activities 'on the spot' by drawing in children's books or having a couple of different tasks with different depths ready. As the structure of the tasks was the same, I could develop many different activities to secure, stretch and deepen pupils' understanding.

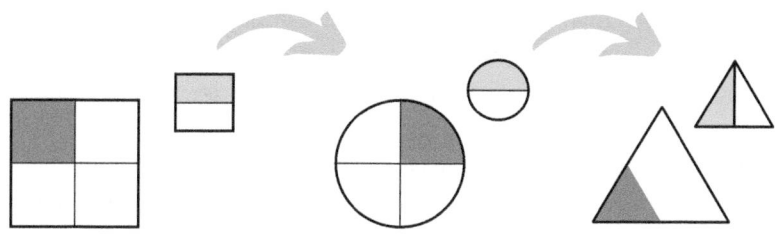

FIGURE 6.11 What is the same? What is different?

The power here is the use of the representations. In this example, the fractions remain the same but the representations are different. Varying the shape and size of the whole but keeping the structure/misconception the same can strengthen understanding to a greater depth. The representations, which we covered in Chapter 3, work with variation theory, and this in turn would work with the stem sentence 'There are ____ equal parts and ____ part is shaded.' Now we're starting to see how these big ideas of teaching to mastery can really come alive.

It is also worth mentioning that designing lessons to include variation theory can reveal what is possible to learn (and not learn), because here we are explicitly teaching a misconception and nipping it in the bud. Here, it is possible to learn that the relative sizes of the whole do not change the fraction that is represented, so children can generalise that $\frac{1}{4}$ is always less than $\frac{1}{2}$. We can then consider what this might look like in children's books and extend it to writing expressions such as: $\frac{1}{4} < \frac{1}{2}$.

Representation	Stem sentence	Expressions
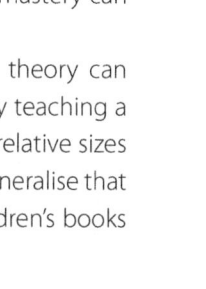	There are ___ equal parts. ___ part is shaded.	$\frac{1}{4} < \frac{1}{2}$
	There are ___ equal parts. ___ part is shaded.	$\frac{1}{4} < \frac{1}{2}$
	There are ___ equal parts. ___ part is shaded.	$\frac{1}{4} < \frac{1}{2}$

FIGURE 6.12 Adding stem sentences to representations

I am sure that there are many other ways in which this could be developed. If you do try implementing something like this in your class or school, please do share it on social media so that we can learn and grow together. Again, when teaching in this way, once this structure is secure, children will learn the structure of the concept. Children may then be able to develop their own examples in their book or they may be able to use the stem sentence structure to compare different fractions, such as $\frac{1}{4} < \frac{1}{3}$ and so $\frac{2}{4} < \frac{2}{3}$ and therefore $\frac{1}{2} < \frac{2}{3}$. You may be surprised at the children's creativity and sophisticated thinking when producing their work.

I mentioned 'thinking on the spot' earlier, but I don't want you to think that I just rock up to lessons and think of what to do there and then. Actually, it's quite the opposite. What I mean is that I will use the same structure and develop this as much I can with children, perhaps encouraging them to explore with different wholes and parts to prove their point. Do be mindful that we can't do this in every lesson or for full lessons, as we'd run out of time very quickly. Think carefully about when this part of a lesson would be useful, and allow it to breathe and develop. The joy of revealing this structure is that the structure will never change. Children have thought about it and attended to it on a deep level, and it will serve them well as they continue to progress through the curriculum and make further links and connections. And to think that I could have given them five minutes to complete some questions from a worksheet. Yikes!

In this example, I have kept the same structure of the original question and changed something small each time. In designing this, I kept on asking myself what I would keep the same and what would be different each time, to draw out the structure. With each small change, we are strengthening children's understanding and allowing them to make connections across the curriculum. Having children draw accurately is important to demonstrate their thinking. I think that this should be encouraged in lessons where possible. I have spoken of my love of the visualiser in lessons, and I would use it to wonder out loud, to explain my thinking and to model the representations that I was drawing with precision. As this sequence of questions progresses, we are able to make connections across questions and link $\frac{1}{2}$ and $\frac{2}{4}$. I am sure that you can look at my example and consider other ways in which it could have been developed and ways in which it could be extended within classrooms. If you can, you're 'thinking on the spot' – adapting, reasoning, explaining and wondering with children. You're pushing beyond answering questions to being genuinely inquisitive about big concepts and explaining why they work. Starting with why and how leads the way to what is possible for children to learn.

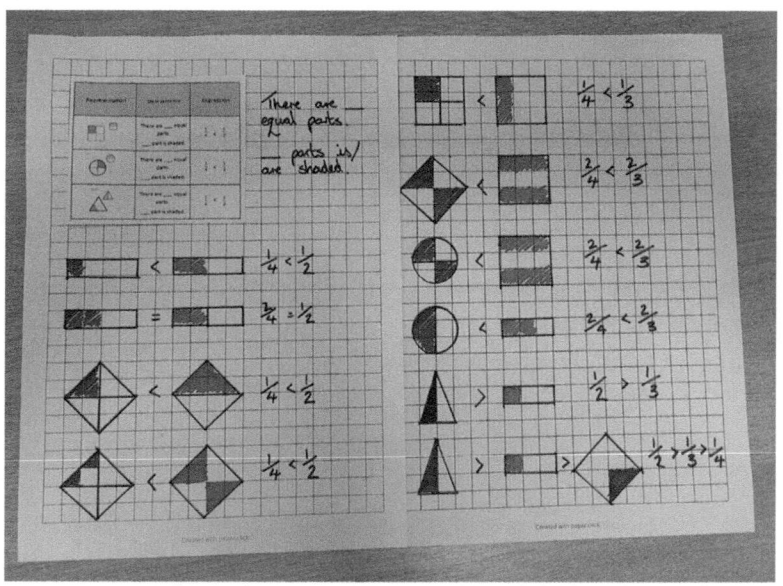

FIGURE 6.13 An example of how children might complete work in their books – notice the deliberate choices made about the representations used.

Conceptual variation: Area and perimeter

Let's consider another big concept: area and perimeter. In the past, I'd have jumped straight in and taught children that perimeter is when you add all of the sides of a 2D shape together, and that to find the area we must multiply length and width together. I look back in horror at how abstract this was for children (and for me), and I would wonder with increasing frustration why children got these two concepts confused. In short, it's because I wasn't teaching it very effectively. After 13 years of teaching, I now have a fair idea of how this concept develops. First, we need to know how this concept builds.

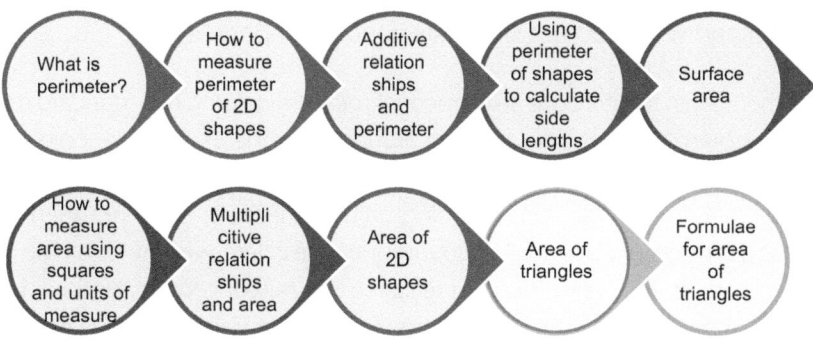

FIGURE 6.14 How the concept of perimeter develops

Area and perimeter are different concepts (although they are connected). Teaching them together or alongside each other, however, can be problematic, as children can become confused.

First, we need to teach children that perimeter is the distance around the edge of a shape. This may be done in many practical contexts. Here, we may ask children to trace the edge of some shapes with their finger to show the perimeter (without measuring anything just yet), and we might develop activities where children have to state what the concept is and is not. On the right-hand-side image in Figure 6.15, we can check understanding by asking whether the dotted line is perimeter or is not perimeter, again linking to the essential features of what perimeter is and is not.

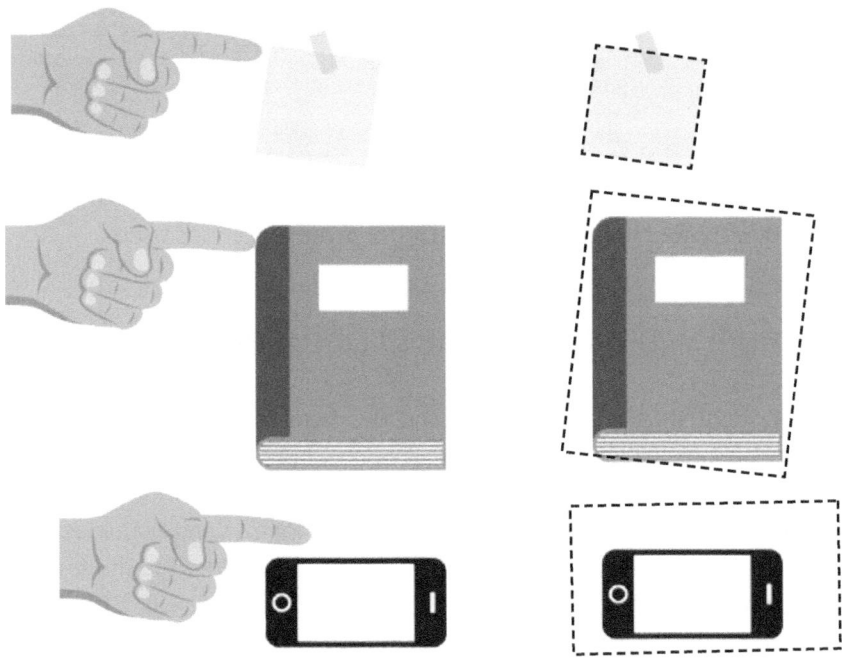

FIGURE 6.15 Developing the concept of perimeter – which show the concept of perimeter and which do not show the concept of perimeter?

We might then begin to introduce an element of measuring, using non-standard measures such as Numicon to measure how many fit around the edge of particular objects, before introducing more formal measuring using a ruler. With both of these examples, we can state the perimeter using the corresponding units of measure. We might then begin to measure the perimeter of 2D shapes by counting squares. Here, we may revisit the concept of what perimeter is and is not by again teaching the misconception. Figure 6.16 shows what I might show children in my maths book under the visualiser or what I might ask children to complete. I will have stuck key vocabulary on 'edges' and perimeter' on the board for children to use in their explanations.

This can then be repeated for different shapes using the squares, progressing to shapes without measurements included and shapes with the perimeter revealed but with some missing measurements. We may then progress to exploring the relationship between opposite sides of regular polygons and multiplication, and the inverse of using division if we know the perimeter and number of equal sides.

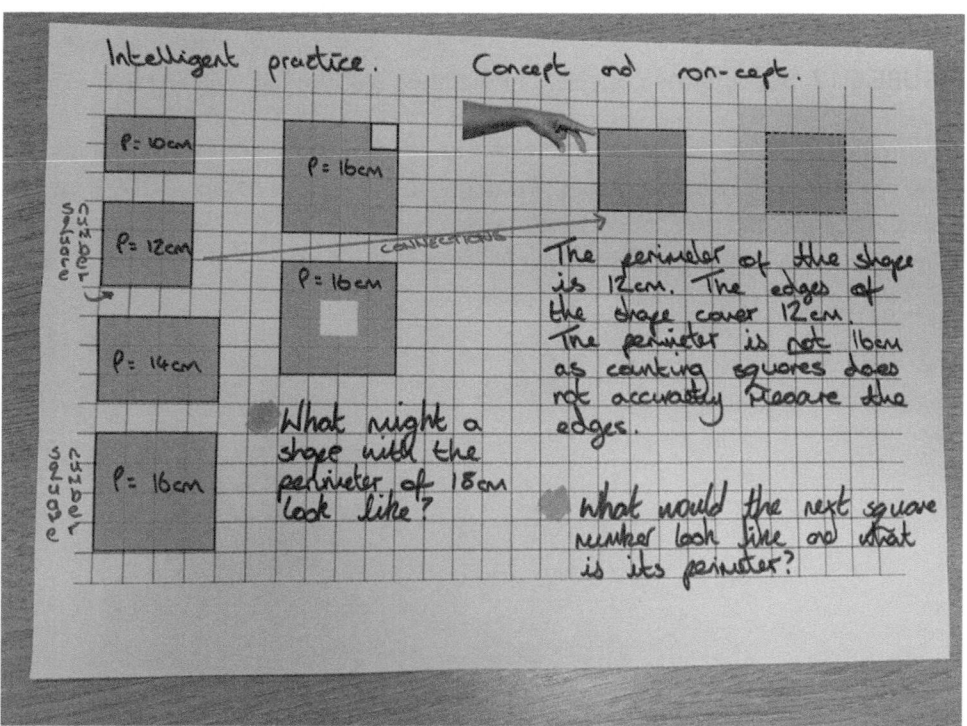

FIGURE 6.16 An example of a task that shows how the concept of perimeter could be practised

For area, we may again begin to build this concept using practical materials. Consider the example in Figure 6.17 overleaf. I am teaching the children about area being the surface of a shape by feeling the surface area and then by covering a textbook in sticky notes. I am showing them what area is. I would also show them what it is not. I could repeat this for different objects until it is clear. I may even ask children to show me a good example (concept) and a bad example (non-cept) to build this idea, using different objects and materials to cover the objects up.

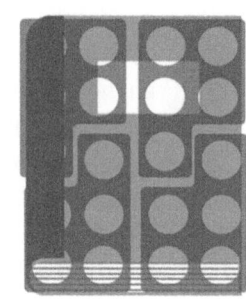

FIGURE 6.17 Building the concept of what area is and what area is not

We can then move to more abstract thinking, using ideas such as shapes as a whole and covering them with different shapes. We might then begin to compare 2D shapes by counting squares and then introducing square centimetres (1 cm²). We might cut out a 1 cm x 1 cm square to illustrate this with children.

FIGURE 6.18 Concrete representation of 1 cm²

We could use these 1 cm² cut-outs to cover 2D shapes on some paper to show how many cm² are covered. This is a favourite activity of mine, as it really brings the concept of area alive (and avoids confusing it with perimeter – after all, children learn what they attend to). We may then progress on to different shapes having the same area by repeating this activity, before formalising it using multiplication with regular shapes, and then introducing length x width = area. We might then introduce composite shapes to apply this understanding and then we can consider what is the same and what is different about the images in Figure 6.19. The highlights (shown circled in the figure below) on the right of the page show where the children could be directed towards developing their thinking using the same structures but at different depths. The last highlight is me 'thinking on the spot', improvising, innovating, creating, designing and playing with the structure of this problem and developing it a little further. Now we know that the area stays the same, I may have even developed this further with a problem like 'show what the greatest perimeter of the shape could be' or 'show what the smallest perimeter of the shape could be'. Thinking on the spot again, I could deepen this problem to include four squares and ask the same questions as before (not to mention the links to a square number), or I might have two 4 cm x 4 cm squares and a rectangle of half that length (2 cm x 4 cm) to deepen and strengthen this concept. The possibilities really are endless, but we need to be very clear about what we want children to learn and draw out from this type of practice.

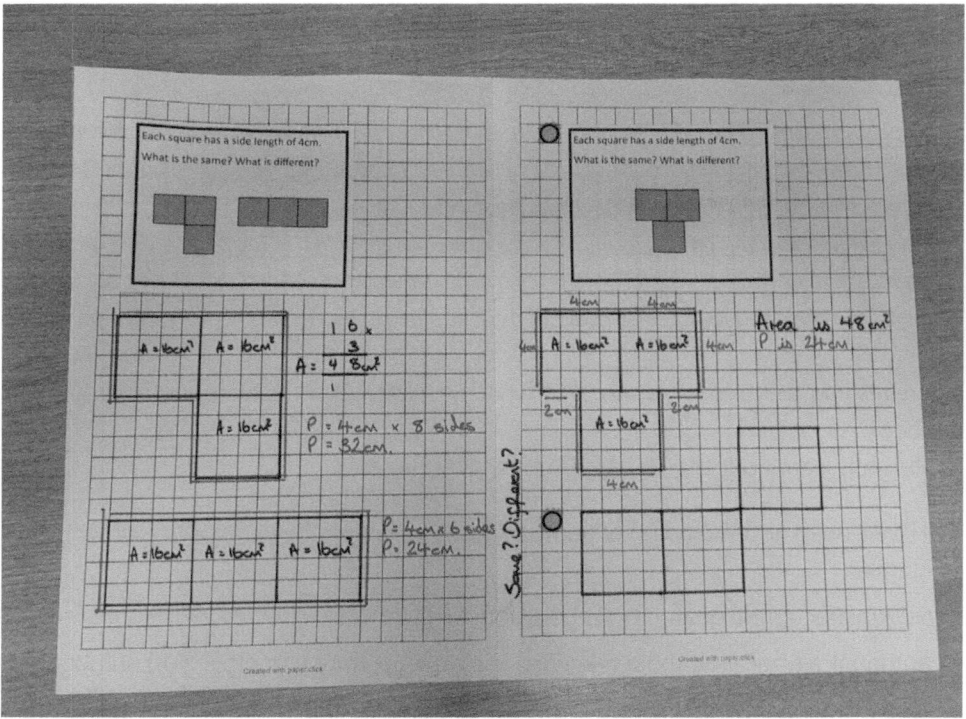

FIGURE 6.19 A possible example of completed work related to area

Early Years

I am sure that there are many Early Years teachers reading this and thinking that they have taught in this way for many years. They'd be right. Many Early Years settings have long used variation theory for early concepts, be it cardinality, counting, comparison, composition, pattern, shape and space, and/or measures. Our youngest children tend to be shown concepts in many different ways. However, these strong foundations are not always built upon across Key Stages 1 and 2 (and beyond). Children are formally introduced to composition from the EYFS, but they are still looking at the composition of numbers well into Year 6 – it's just that this time they are looking at numbers to ten million or decimal fraction numbers.

The statutory framework for Early Years states:

'Developing a strong grounding in number is essential so that all children develop the necessary building blocks to excel mathematically. Children should be able to count confidently, develop a deep understanding of the numbers to 10, the relationships between them and the patterns within those numbers. By providing frequent and varied opportunities to build and apply this understanding – such as using manipulatives, including small pebbles and tens frames for organising counting – children will develop a secure base of knowledge and vocabulary from which mastery of mathematics is built. In addition, it is important that the curriculum includes rich opportunities for children to develop their spatial reasoning skills across all areas of mathematics including shape, space and measures. It is important that children develop positive attitudes and interests in mathematics, look for patterns and relationships, spot connections, "have a go", talk to adults and peers about what they notice and not be afraid to make mistakes.' (DfE, 2023, pp. 10–11)

It is refreshing that it mirrors much of what we have been thinking about throughout this book.

The early learning goals (ELGs) for number are:

- 'Have a deep understanding of numbers to 10, including the composition of each number.
- Subitise (recognise quantities without counting) up to 5.
- Automatically recall (without reference to rhymes, counting or other aids) number bonds up to 5 (including subtraction facts) and some number bonds to 10, including double facts.' (DfE, 2023, p. 14)

So, we might start by introducing children to the idea of all (whole) numbers being made of 1s and focus on their composition. We might then look at 'parts' of the 'whole'. We're still using this language well into Key Stage 2. We might then focus on comparison using 'more than', 'fewer than' and 'equal to', by making equal and unequal sets and asking whether they are equal or not equal (concept and non-cept). We may explore odd and even numbers and ensure that children are secure with the composition of numbers to 10.

Composition progression

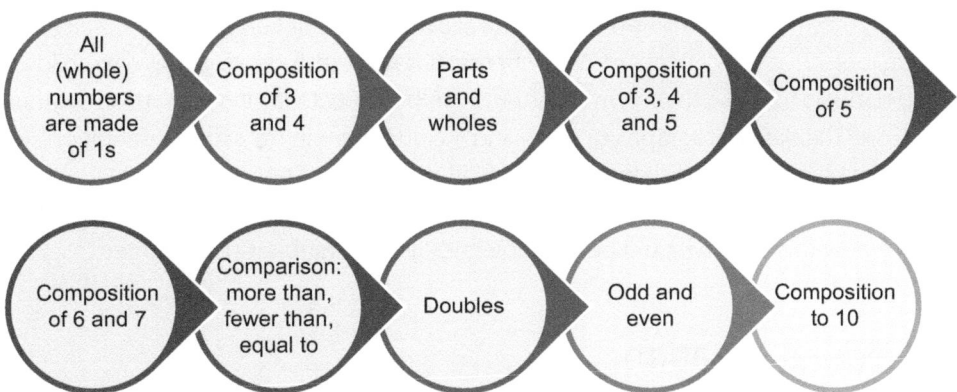

FIGURE 6.20 Progression of composition in Early Years

Let's look at the composition of 7 in particular. Imagine that we have done lots of work on securing the composition of 5 and 6, looking at the parts that make 5 and having different ways to make 5 in our provision. We might build upon this understanding to show the concept of 7 in lots of different ways. Here we could play different games to implement this, which allows you to use variation in the representation to show different compositions.

How do you see it?

Using dot patterns or Hungarian number frames, we might ask children to say how many flash on the screen and how they saw it. The examples in Figure 6.21 are not extensive but are there to give an example of the use of shading/colour in representing the concept of 7 and in changes in formation of the 7. Take note of the way in which subitising different numbers based on their colour could work too (and the fact that there can be more than two parts to 7).

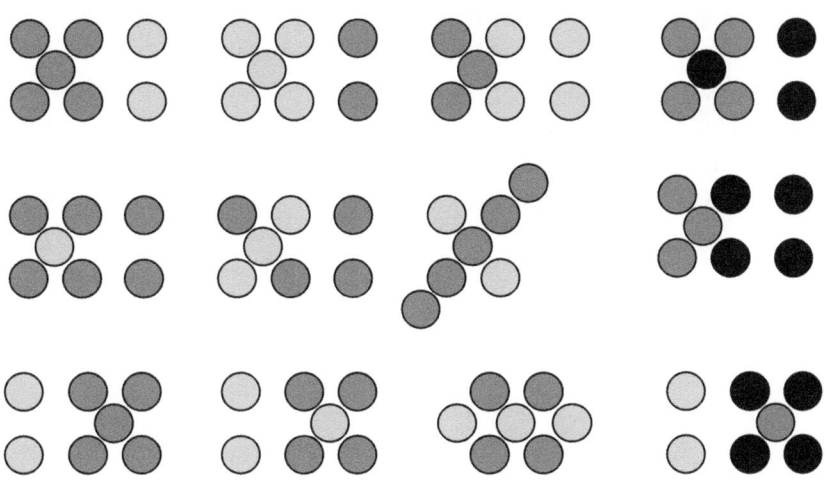

FIGURE 6.21 Dice patterns (Hungarian number squares) used to subitise

7 or not 7

This game can be adapted for any number within 10. Using the examples we have just looked at they might flash on the screen one at a time and children state whether it is 7 (if you are looking at the concept of 7) or whether it is not 7 (the non-cept). Giving children bingo dabbers to make their own 7 patterns can be a great way in which to strengthen this. I used to take photographs of them and include them in my slide presentations (or take them on a tablet and slide them along on the screen). Stating that 5 and 2 make 7 and writing the expression can support children's understanding too. I hope that you can also see how this activity could be done practically with double-sided counters.

Procedural variation

We have explored conceptual variation in some depth in this chapter. We now turn to procedural variation. For me, this is something to which I came back again and again. I'd never have a 'variation theory lesson' – rather, small activities within lessons that applied variation theory to draw out patterns, connections and relationships. This section will give practical examples of what this looks like and how to implement it into your practice.

For me, procedural variation has been a game-changer. I first came across its use in Shanghai, when I visited in 2019. This approach generally relates to calculation and number, and tends to keep something the same and something different. It allows learners to make meaningful connections by highlighting the variations. In doing so, and with careful thought, we can reveal the deep mathematical structures that are at play.

Before we look at how we might implement, embed or sustain this in our own classrooms or across school(s), take a look at the questions in Table 6.1.

What strikes you about each set of the sequences? What do you think that I am intending to draw your attention to? At first glance, they may appear to be very procedural, and perhaps some teachers may consider them 'too easy'. The beauty here is the potential of where these sequences can go and what patterns and structures we can get from them. In the addition example, we may generalise that if we add one to an addend, we add one to the sum. Notice that there is a clear process to get children looking for patterns and connections, and then something changes. This creates a small cognitive shock and means that the activity does not become a straightforward 'fill in the gaps' but develops into a rich and deep sequence.

TABLE 6.1 Procedural variation with calculation

Addition	Subtraction	Multiplication	Division
25 + 5	25 – 5	5 x 2	24 ÷ 4
25 + 6	25 – 6	2 x 5	28 ÷ 4
25 + 7	25 – 7	4 x 5	32 ÷ 4
25 + 8	25 – 8	2 x 10	36 ÷ 4
35 + 9	35 – 9	2 x 2 x 5	360 ÷ 4

When teaching a method or procedure, we can modify it from one question to the next while keeping a part of it consistent. Children then build on their previous knowledge and apply it to the new question.

Multiplication times tables and procedural variation

Let's take a look at a Key Stage 1 example. Imagine that you are teaching children about the 10 times table and you've looked at groups of 10, taught them about factors and products and counted forwards and backwards in 10s. Now you're ready to set children off with a task. You could give children a list of calculations to solve, such as:

$1 \times 10 =$ $\qquad$ $2 \times 10 =$ $\qquad$ $3 \times 10 = \ldots$ $\qquad$ and so on.

Or we could think carefully and deliberately about a careful sequence of numbers that will eventually lead children towards making a generalisation. In Figure 6.22, we might ask children: 'What is the same? What is different?'

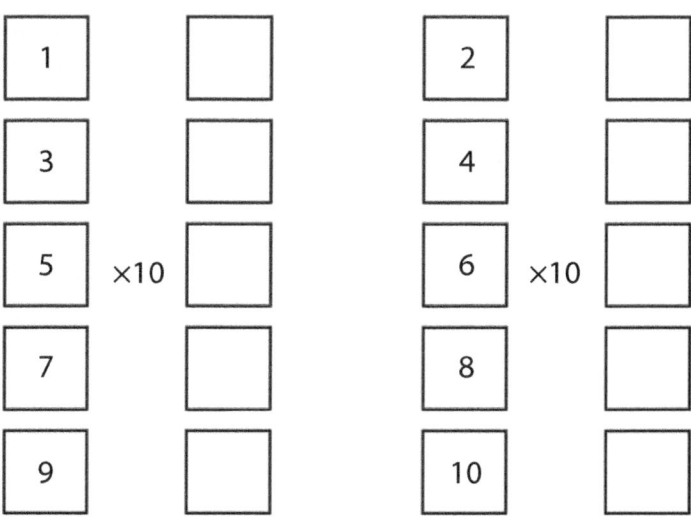

FIGURE 6.22 What do we want children to notice?

Here, I want to draw attention to the fact that the factors are grouped into odd and even numbers. I might ask children to begin to populate the missing numbers and then ask what they notice. We would then draw attention to odd x even = even and even x even = even. Once this is secure, we might ask children whether the answer to something like 19 x 10 would be odd or even, or even something like 112 x 8. In doing so, children can apply the generalisation that they have found and continue to apply it again and again and again.

We can see how children in Key Stage 1 might thus be able to reason about the types of answers in the questions in Figure 6.23, using the generalisation. This time, the task for children may be to write 'odd' or 'even' in the product box. Notice the slight variation in the factor box too (here, I might check children's understanding of numbers between the factor before and after in the sequence). To really personalise this, I might finish with asking children to create their own questions to give their friends in class.

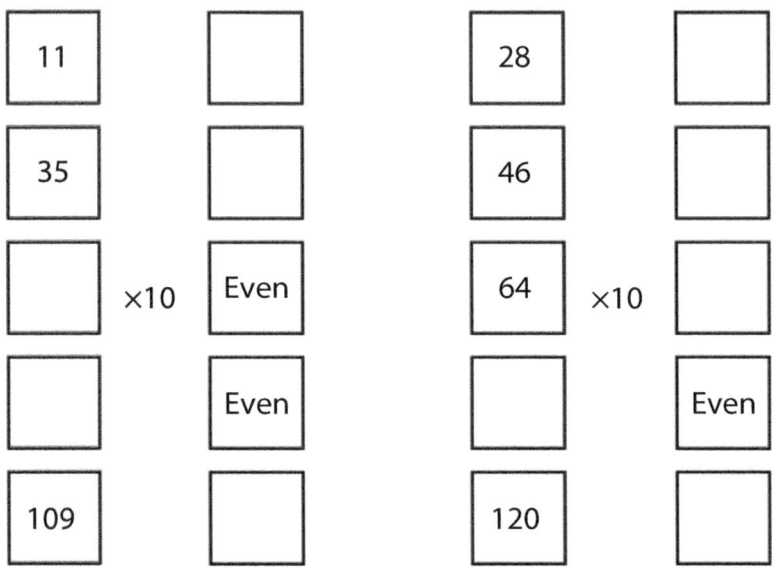

FIGURE 6.23 What do we want children to notice?

The end of this small step or series of sessions might result in me asking: 'What type of product does an odd x odd expression give us?' We might, though, decide that we want to have children compare expressions using the equal and inequality symbols. Take a look at both sequences in Table 6.2. What do you notice? What is the same? What is different? Which set do you think is most effective?

TABLE 6.2 Comparing procedural variation with calculation

A	B
10 x 10 ☐ 1 x 10	10 x 10 ☐ 9 x 10
5 x 10 ☐ 2 x 10	10 x 10 ☐ 9 x 10 + 10
10 x 5 ☐ 3 x 10	10 x 10 ☐ 10 x 10 + 10
7 x 10 ☐ 1 x 10	10 x 10 ☐ 11 x 10 – 10
11 x 10 ☐ 4 x 10	10 x 10 ☐ 12 x 10 – 20
2 x 10 ☐ 4 x 10	10 x 10 ☐ 24 x 10 – 120
6 x 10 ☐ 9 x 10	10 x 11 ☐ 24 x 10 – 110

In Table 6.2, Set A is what I used to do: teach the skill and give children random practice. Set B is how I now structure questions. Notice how the left-hand factors in set B (almost) stay the same each time, to anchor children's understanding, and the slight changes that are made each time to the expression on the right-hand side.

I want children to notice the patterns in the expressions; however, I do not want children to simply mindlessly complete these patterns in books. This would mean children are not spending time thinking about the structure of the questions. Notice the second-to-last question in set B. It's meant to act as a cognitive shock so that children need to think deeply, but look closely at the connections to be made with the numbers chosen.

We can see how this way of sequencing can progress to even deeper thinking in Key Stage 2. I had mentioned previously about 'thinking on the spot'. On reflection, this is probably a poor phrase; I'm thinking deeply about the structures, where this use of variation can take the children and what I want them to notice. Take the Key Stage 2 example in Figure 6.24. This is one example.

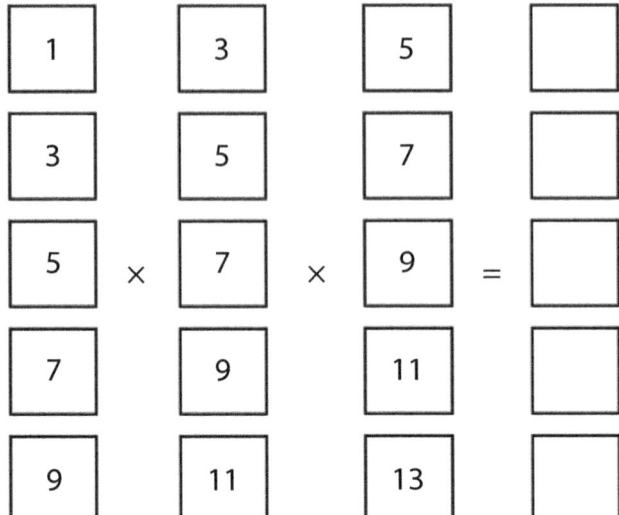

FIGURE 6.24 What could children generalise about the types of factors and products that they will obtain?

We could have varied it with the factor being a missing part and some products being included. It's not then a giant leap to pose questions such as: 'What if one of the columns had even factors?' Again, we're starting to draw attention to generalising odd x odd x odd = odd and odd x odd x even = even, and so then we can move to odd x even x odd = even and even x even x odd = even. We're now revising commutativity. Personalising in this way (or, dare I say it, 'differentiating') can lead to deep understanding of structure. It's also clear how this links to other areas of the maths curriculum, such as volume.

Teaching with procedural variation can be applied across much of the curriculum. The rest of this chapter provides some suggestions. I'm sure that you're starting to get the impression that there may be infinite ways in which to apply this in your teaching. If you are using a published scheme, it is easy to miss this. Some schemes use variation theory sporadically, and it is not always clear what the thinking is around this. A simple way around it can be to use the first question suggested (if it's any of the four operations or anything linked to number, including place value or fractions) and then consider the learning objective and how you might develop a sequence of questions that create patterns, links and connections. In designing questions like this, they end up having an element of reasoning designed into them.

Place value

One of the key objectives for children is to understand the place value of each digit in two-digit numbers by the end of Key Stage 1.

We might be tempted to give children a mixture of random two-digit numbers and then move one if they can do it, but there's power in developing a careful sequence to supercharge thinking about 10s and 1s. Prefix the numbers below by asking: What is the same? What is different?

- 12 and 21
- 23 and 32
- 34 and 43

> One 10 and two 1s (12)
> Two 10s and one 1 (21)

In Key Stage 2, we might ask children the same questions for decimal numbers:

- 1.2 and 2.1
- 2.3 and 3.2
- 3.4 and 4.3 (or even 34 and 3.4)

> One 1 and two tenths (1.2)
> Two 1s and one tenth (2.1)

Vocabulary is key to the explanation and thinking here (1s, tenths, part, whole) and what you want to draw attention to.

Addition and subtraction

We return to a familiar representation of part–whole models to explore the structure of addition and subtraction. Again, we're going to avoid random calculations and this time look at related facts and powers of 10.

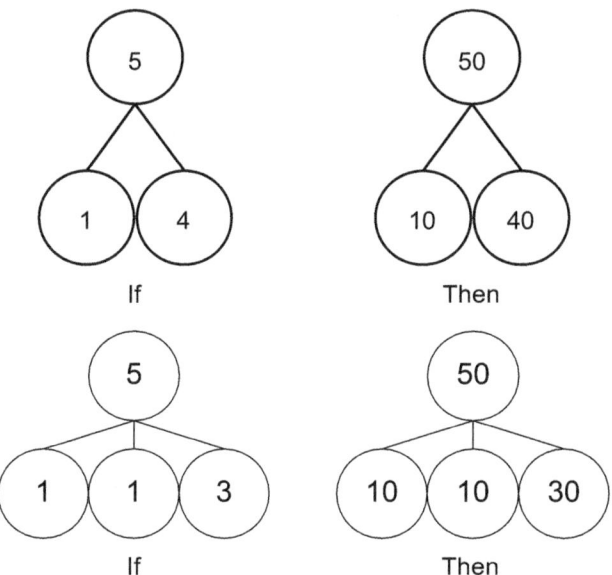

FIGURE 6.25 Making connections

We might even extend this to allow children to explore the related facts of addition and subtraction. It really does depend on what you want children to notice and learn.

Here we are drawing attention to the same digits having a different value, based on their position within the number. We might explore this using dienes and rods or part–whole models to show how this works.

In Chapter 3, we showed how this can be done effectively using a number line. Using procedural variation allows this idea to develop further and make connections.

Multiplication and division

In the example of the 5 times table in Table 6.3, I want children to use the structures to help them to answer the next question without counting on in their head or using elaborate methods.

TABLE 6.3 Procedural variation with calculation

Use >, < or =		Use >, < or =		Use >, < or =	
2 x 5	5 x 2	2 x 5	4 x 5	5 x 5	10 x 5
2 x 5	5 x 2 + 5	2 x 5	8 x 5	5 x 10	10 x 5
2 x 5	5 x 3	2 x 5	16 x 5	5 x 5 + 5	10 x 5
2 x 5	5 x 6	2 x 5 x 2	16 x 5	5 x 5 + 25	10 x 5
2 x 5	5 x 6 – 5 x 4	5 x 2 x 2	5 x 16	5 x 10 + 5	11 x 5

Take a look at the sequence below. I have included the answers within it. Can you see the thread of thinking and the link between each question so that children can make connections?

- 5 x 2
- 2 x 5
- 2 x 6

- 2 x 7
- 4 x 7
- 4 x 8
- 4 x 9
- 4 x 3
- 5 x 3
- 6 x 3
- 12 x 3
- 12 x 6
- 12 x 12
- 24 x 12

These sequences can take time to think about, in order to consider the links and connections that can be made. They can make a great staff meeting to begin the process of thinking about procedural variation. I am sure that you can think of many other ways in which my sequence could have been developed.

This thinking can then be extended to division. What generalisation do you think that I am trying to draw attention to in the sequences in Table 6.4?

TABLE 6.4 Making connections with division

24 ÷ 2 = 12	48 ÷ 2 = 24	96 ÷ 2 = 48
24 ÷ 4 = 6	48 ÷ 4 = 12	96 ÷ 4 = 24
24 ÷ 8 = 3	48 ÷ 8 = 6	96 ÷ 8 = 12

If I work downwards, I can generalise that if I double the divisor and keep the dividend the same, the quotient (answer) halves. I can also work horizontally to make connections that if the divisor is doubled and the dividend is the same, the quotient doubles.

Fractions

As with most areas of the curriculum and procedural variation, we can take this in many different directions. In the example in Table 6.5, I want children to spot a pattern in the first few questions, but then I want them thinking deeply about the connections between the numbers used and the underlying structure of finding $\frac{1}{4}$ of numbers. Here, children will generalise that the greater the whole, the greater the parts will be if the fractions remain the same.

TABLE 6.5 Opportunities to develop beautifully connected sequences of questions using procedural variation

Doubling pattern	Use >, < or =		Use >, < or =	
$\frac{1}{4}$ of 4	$\frac{1}{4}$ of 4 ☐ $\frac{1}{4}$ of 8		$\frac{1}{4}$ of 4×5 ☐ $\frac{1}{4}$ of 5×4	
$\frac{1}{4}$ of 8	$\frac{1}{4}$ of 40 ☐ $\frac{1}{4}$ of 80		$\frac{1}{4}$ of 4×5 ☐ $\frac{1}{4}$ of 8×5	
$\frac{1}{4}$ of 16	$\frac{1}{4}$ of 40 ☐ $\frac{1}{4}$ of 44		$\frac{1}{4}$ of 4×5 ☐ $\frac{1}{4}$ of 2×5	
$\frac{1}{4}$ of 32	$\frac{1}{4}$ of 40 ☐ $\frac{1}{4}$ of 36		$\frac{1}{4}$ of 4×5 ☐ $\frac{1}{4}$ of $2 \times 5 \times 2$	
$\frac{1}{4}$ of 64	$\frac{1}{4}$ of 40 ☐ $\frac{1}{4}$ of 52		$\frac{1}{4}$ of 4×5 ☐ $\frac{1}{4}$ of $2 \times 5 + 10$	

Now of course, I don't do this in every single part of a lesson, but I have had lots of success when adding procedural variation to sequences of questions in lessons. I'd always model the first or even second question to explore what was the same and what was different. I would draw attention to links and connections. I would also cover up the subsequent questions and only reveal these one at a time, asking children to predict what the next question could be. Once modelled, I'd give children some similar questions to complete in their books.

Applying this technique in Key Stage 2 works in much the same way. I could use any year group and apply this thinking in order to carefully design questions that allow intelligent practice.

Place value – Year 6

Let's use Year 6 as an example for now. Look at how the digits remain the same but there is a change in their position, and so a change in their place value.

TABLE 6.6 Opportunities to develop beautifully connected sequences of questions using procedural variation

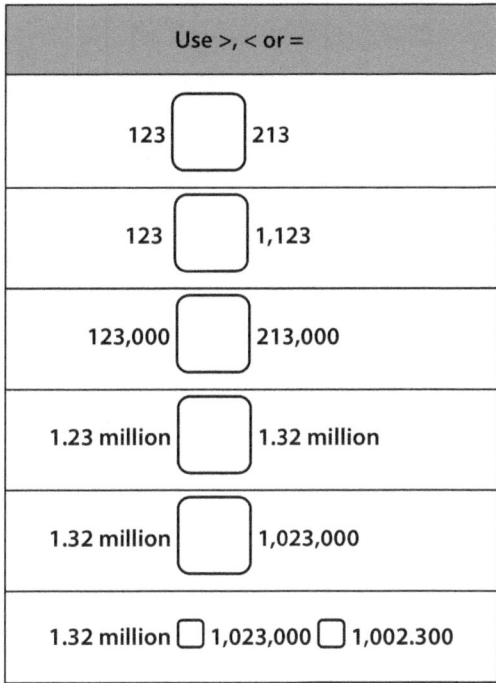

Use >, < or =
123 ☐ 213
123 ☐ 1,123
123,000 ☐ 213,000
1.23 million ☐ 1.32 million
1.32 million ☐ 1,023,000
1.32 million ☐ 1,023,000 ☐ 1,002.300

Fractions – Year 6

I have written about 'tracking back' in maths in this book. Using procedural variation in many areas of maths can allow us to start sequences of questions at a very simple starting point, so that all children can access them before moving on to more complex questioning, following a clear and careful thread of thinking to allow children to make connections and links. In the following sequence for Year 6, I start with something that most should be able to do, and I change the sequence each time only slightly. Glance at Table 6.7 and see whether you can follow my train of thought for each of the focus areas.

TABLE 6.7 Opportunities to develop beautifully connected sequences of questions using procedural variation with fractions

Adding fractions	Multiplying fractions	Fractions of an amount
$\frac{1}{4}+\frac{1}{4}$	$\frac{1}{2}\times 2$	$\frac{1}{2}$ of 6
$\frac{1}{4}+\frac{2}{4}$	$\frac{1}{2}\times 3$	$\frac{1}{2}$ of 60
$\frac{1}{4}+\frac{1}{2}$	$4\times\frac{1}{2}$	$\frac{1}{2}\times 66$
$\frac{1}{4}+\frac{3}{4}$	$\frac{1}{3}\times 2$	$\frac{1}{2}$ of 6.6
$\frac{1}{4}+\frac{4}{4}$	$\frac{2}{3}\times 2$	$\frac{1}{2}$ of 6.66
$\frac{1}{4}+\frac{5}{4}$	$\frac{4}{3}\times 2$	$\frac{1}{3}$ of 6.66
$\frac{1}{4}+\frac{6}{4}$	$3\times\frac{4}{3}$	$\frac{1}{6}\times 6.66$
$\frac{1}{4}+1\frac{2}{4}$	$4\times\frac{4}{3}$	$\frac{1}{3}$ of 666
$\frac{1}{4}+1\frac{1}{2}$	$8\times 1\frac{1}{3}$	$\frac{1}{3}\times 0.6$
$\frac{1}{4}+\frac{3}{6}$	$8\times 1\frac{2}{3}$	$\frac{1}{3}$ of 0.06
$\frac{2}{4}+1\frac{2}{4}$	$8\times 2\frac{1}{3}$	$\frac{1}{6}$ of 0.06

What if we were teaching children something like $1-\frac{3}{4}$. How could we use variation theory to support thinking around this?

We might logically start by providing children with examples of wholes, such as $1=\frac{2}{2}$, $\frac{3}{3},\frac{4}{4},\frac{5}{5},\frac{10}{10},\frac{100}{100}$ and then generalising that when the numerator and denominator are equal, the fraction is equivalent to one whole. We may then have children think about $1-\frac{3}{4}$ as $\frac{4}{4}-\frac{3}{4}$. Once this is secure, we can begin to create beautiful sequences to strengthen this thinking. Perhaps now would be a good time for you to start with this calculation and develop a series of questions that all link to each other. I've included one example of how this *could* be developed below.

- $1 - \dfrac{3}{4}$

- $1 - \dfrac{2}{4}$

- $1 - \dfrac{1}{4}$

- $1 - \dfrac{1}{5}$

- $1 - \dfrac{2}{5}$

- $2 - \dfrac{2}{5}$

- $3 - \dfrac{2}{5}$

- $3 - \dfrac{4}{5}$

- $3 - \dfrac{6}{5}$

- $3 - 1\dfrac{1}{5}$

- $3 - 2\dfrac{1}{5}$

For now, have a look at something that you have taught or something that is coming up. Have a look at a procedural question and scribble it down. Consider how you might develop the sequences of questions with a common thread through them so that children can make connections. In most cases, the relationship that you want children to notice needs to be explicitly taught. For the example 3 x 5 and 6 x 5, giving children the generalisation of 'double the factor, double the product' means that children can then apply this to many areas of maths.

Use the space in Table 6.8 to scribble down an initial question. Perhaps share it with staff and see all of the wonderful ways in which you can take this further. See how you might go back and edit those sequences to include other small connections and lead children to deep and rich understanding of maths.

TABLE 6.8 Develop your own beautiful sequences using procedural variation

Addition	Subtraction	Multiplication	Division	Fractions	Decimals

Implementing this in your own classroom takes time but is a lot of fun. Having staff meetings that centre on careful sequences of variation, focusing on what it makes us attend to and learn and how we use prior knowledge to inform our thinking for the next question, is time well spent. Once embedded, I have found that this approach makes children think differently. They start to look for patterns, connections and links and begin to realise that maths is not about unrelated, abstract facts but that everything is connected. Teaching with variation transformed my teaching and my thinking. It's powerful stuff.

Key takeaways

Variation plays a critical role in teaching when teaching for mastery. Done well, it has the power to transform teaching and develop deep and lasting mathematical understanding in students. Variation theory, as discussed, is an instructional strategy that helps students to discern and understand mathematical concepts by focusing on both the essential and non-essential features of these concepts.

One of the key points highlighted in the chapter is the importance of conceptual variation. This approach helps students to understand what a concept is and what it is not, by comparing various examples and non-examples. For instance, using different representations to teach the concept of fractions enables students to see that the size of the parts can vary while the fraction remains the same, thus deepening their understanding of the relative sizes of fractions.

By using deliberate representations that expose what a concept is and is not, and using carefully designed sequences of problems, teachers can guide students to see underlying patterns and structures. For example, varying the context in which a fraction is presented while maintaining the same structure helps students to generalise the concept of fractions beyond a single representation.

This chapter underscores the value of adapting lessons in the moment, using variation theory to respond to students' needs in real-time. This adaptive teaching approach ensures that lessons are personalised for all learners and that children are constantly challenged to deepen their understanding. This method allows for immediate adjustments to teaching strategies, thereby fostering a more dynamic and responsive learning environment.

While this approach is powerful, it requires careful planning and thought. Teachers must be deliberate in their choice of examples and non-examples and in the sequence of the problems that they present. This careful planning ensures that students can make meaningful connections and develop a robust understanding of mathematical concepts.

In transitioning to the next chapter on mathematical vocabulary, it is important to note that the precise language used in mathematics is integral to students' understanding. Just as variation helps to clarify concepts by focusing on essential features, a strong grasp of mathematical vocabulary allows students to articulate their understanding clearly and accurately. The upcoming chapter will delve into strategies for teaching mathematical vocabulary effectively, ensuring that students are equipped with the language skills necessary to describe and discuss mathematical ideas with precision. This linkage underscores the interconnected nature of mathematical teaching practices, where variation and, indeed, all of the other ideas of teaching for mastery work hand in hand to build a comprehensive and profound mathematical foundation for students.

7 Vocabulary

Like every subject area, maths has its own unique vocabulary. A key theme of this book is leaving nothing left to chance. Vocabulary should not be left to chance either. It needs to be deliberately planned for and systematically used in lessons. A key barrier to doing this can be teacher subject knowledge. I spent a lot of my teaching career explaining concepts with imprecise vocabulary. I'm sure that many of us have taught lessons where explanations have been underdeveloped and reasoning is vague. Vocabulary can go a long way to offset that.

In phonics, we expect children to understand technical vocabulary of split digraphs, graphemes, phonemes, segmenting, consonant digraphs and vowels. As teachers, we systematically plan the explicit teaching of these words and these concepts. Maths should be much the same. Knowing technical mathematical vocabulary can unlock a world of reasoning and gives children the tools to articulate their thinking. It can turn standard activities linked to calculation into reasoning activities that allow children to make rather impressive mathematical generalisations. Implementing this in school can be an 'easy win', but only if colleagues see the value in teaching the technical vocabulary and how it can unlock reasoning without limits.

Addition

How would we describe the following calculation to children?

$1 + 2 = 3$

We might point at the digits and show children which number we are referring to, or we may simply read the calculation out to children.

In this example, 1 is an addend (sometimes the first number in an addition is called the augend), 2 is an addend and 3 is the sum (answer). Therefore 'sum' is the answer of addends combined together. This can have huge implications on how we teach. How many of us were told to 'do our sums' at school and were met with a mixture of different operations to complete? It's easy to see how misconceptions can creep into thinking. If we explain to children, from a young age, what each of the 'parts' of the calculations are, we can unlock reasoning. Let's take the example we have used and apply some procedural variation to the next set of calculations: $2 + 2 = 4$; $3 + 2 = 5$; $4 + 2 = 6$; $5 + 2 = 7$. Here, children can see that the impact of adding 1 to an addend is that it adds 1 to the sum. In exposing this relationship, children start to make connections and think mathematically. This can lead to further generalisations, that addends can be added in any order and so are commutative.

Moreover, children can then apply this rule to any further concepts relating to addition as they progress through school.

Now that teachers know this, we can apply it to all areas of mathematics that require addition.

- $0.1 + 0.2 = 0.3$ (addend + addend = sum)
- $0.03 = 0.01 + 0.02$ (sum = addend + addend)
- $\frac{1}{10} + \frac{2}{10} + \frac{3}{10} = \frac{6}{10}$ (addend + addend + addend = sum)

Even though this vocabulary can be taught from Year 1 when teaching addition, children will revisit addition throughout their education and so their reasoning will be much tighter. The commutative nature of addition can therefore be explained using algebra.

$a + b = c$ (a is an addend, b is an addend and c is the sum), so $b + a = c$ (and so $c = a + b$ and $c = b + a$).

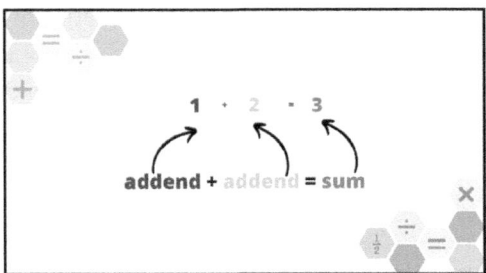

FIGURE 7.1 The vocabulary of addition

Subtraction

Imagine that we are teaching $3 - 2 = 1$ (this could be $30 - 20 = 10$ or $0.3 - 0.2 = 0.1$ in other year groups). Which vocabulary would support children here?

In this calculation, 3 is the minuend (the total amount), 2 is the subtrahend (the amount to be taken away from the minuend) and 1 is the **difference**. Experience tells me that some children still find the concept of difference confusing well into Key Stage 2 and Key Stage 3. When it is associated with subtraction systematically and deliberately from a young age, it is clear to see how this can support children's understanding of the concept of difference. Again, when children are subtracting in lessons, the use of this vocabulary allows for sharply focused reasoning, which can be applied to column subtraction. We must, however, explicitly teach children the structure of subtraction. Using representations can reveal this structure.

Minuend	
Subtrahend	Difference

Minuend	
Difference	Subtrahend

FIGURE 7.2 The vocabulary of subtraction

Here, we can teach children that if we swap the values of the subtrahend and the difference, the minuend remains the same: 3 − 2 = 1 so 3 − 1 = 2.

3	
2	1

3	
1	2

FIGURE 7.3 Connecting the vocabulary to the representation to reveal the structure

Notice the subtle change in the relative size of the parts for the subtrahend and difference, so that children do not develop a misconception that they are equivalent.

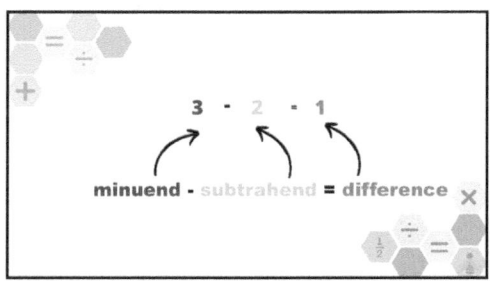

FIGURE 7.4 The vocabulary of subtraction

Multiplication

Teachers may be most familiar with the vocabulary linked to multiplication. How would you describe 3 x 2 = 6 to a class? In this case, 3 is a factor, 2 is a factor and 6 is the product.

My experience of many years teaching in Year 6 is that children often look at these words as though they have never been uttered out loud before, never mind taught in different year groups. The power of using this vocabulary with children allows them to make the generalisation that when factors are multiplied they create a product (factor x factor = product and so factor x factor x factor = product). We can also allow children to make connections – for example, 3 x 2 = 6 so 6 x 2 = 12 – leading to the generalisation of 'if you double one factor, the product doubles'.

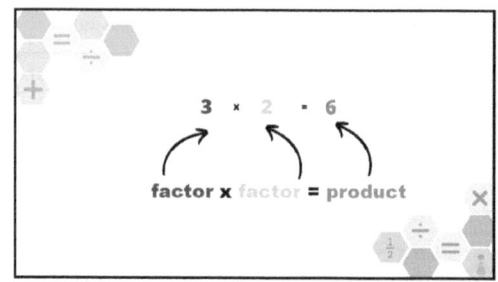

FIGURE 7.5 The vocabulary of multiplication

Division

For division, the use of technical vocabulary works in much the same way as we have described for the other operations. We explored how we can turn what seem like routine fluency questions into supercharged reasoning questions in Chapter 6.

Let's take the example $10 \div 2 = 5$. In this, 10 is the dividend (the amount to be divided), 2 is the divisor (the number by which to divide the divisor) and 5 is the quotient (the answer to a division).

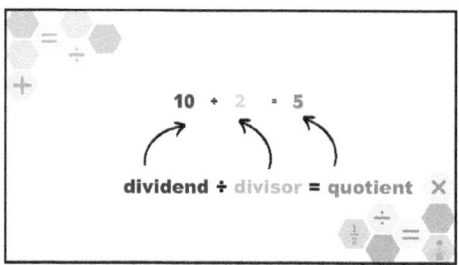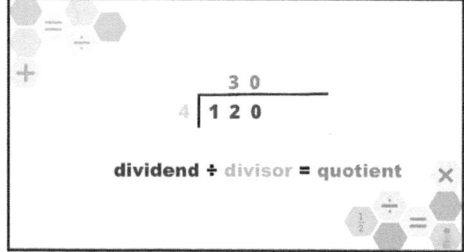

FIGURE 7.6 The vocabulary of division

We can then consider the structure of division and the impact of swapping the divisor and quotient.

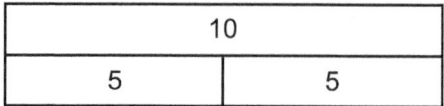

FIGURE 7.7 The structure of division when swapping the divisor and quotient

Here we can see that the 10 (dividend) divided by 2 (subtrahend) gives us a quotient of 5 but when the divisor and quotient are swapped to 10 (dividend) divided by 5 (divisor) and 2 (quotient) the structure of the division changes. We can see how the first bar model in Figure 7.7 uses the multiplication structure of 5 two times ($2 \times 5 = 10$) and in the second bar model we have 2 five times (5×2). This can be expressed algebraically: $a \div b = c$ and so $a \div c = b$.

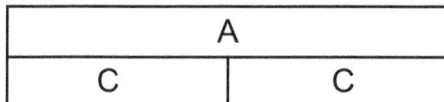

FIGURE 7.8 The structure of division when swapping the divisor and quotient

Implementing this technical vocabulary across the curriculum can unlock reasoning without limits. It gives children the tools with which to reason about number with higher degrees of accuracy. In turn, this allows children to think mathematically.

Let's look at how this vocabulary might then support children in noticing generalisations. We could tell children, for example, $136 \div 4 = 34$. We might reasonably teach children that 134 is the dividend, 4 is the divisor and 34 is the quotient. This could be done by teaching them short division.

We might then apply procedural variation theory and ask:

- $137 \div 4 = 34$ remainder 1
- $138 \div 4 = 34$ remainder 2
- $139 \div 4 = 34$ remainder 3
- $140 \div 4 = 35$

The reasoning that comes from this can be great when children are familiar with the language of division. Children may recognise that the remainder can never be greater than or equal to the divisor. Without this specific vocabulary, reasoning may become vague and clumsy. Using the example from above, we might then ask children questions such as: 'Would $400 \div 4$ have a remainder? So would $401 \div 4$ have a remainder?' This allows children to make generalisations, predictions and connections. They can then check using short division. This is a great way of demonstrating how the big ideas of teaching for mastery work together to inform our pedagogy.

We may also draw attention to the relationship between multiplication and division. If $5 \times 2 = 10$ then $10 \div 2 = 5$ and so $10 \div 5 = 2$. Here, the value of the product in the multiplication becomes the dividend. The divisor and quotient can swap positions. This often comes up when teaching inverse operations. Using this vocabulary makes it crystal clear.

Having a shared vocabulary in school is important in maths (and most other subjects), but this may be particularly true for fractions. Lots of children (and adults) can find the concept of fractions difficult. The way in which they are taught, alongside the language used with fractions, supports how much children understand them.

Fractions

In this section, I have put some of the key vocabulary associated with fractions in bold to draw attention to how this can become almost otherworldly for children if it is not taught in a careful and coherent way.

I can remember meeting my new Year 6 class a few years back and asking them what a **fraction** was. They responded with $\frac{1}{2}$ and $\frac{1}{4}$. I told them that these were examples of a fraction but asked again what a fraction was. I was met with confused faces. This told me two things: children had encountered fractions in their education but they may have some misconceptions about them. I needed to start at the beginning. I told them that a fraction is a **part** of a **whole**. The vocabulary of 'part' and 'whole' was useful here, as they had heard it for years when partitioning numbers in Key Stage 1. To build a working concept of fractions, I began with decontextualised stem sentences and images, such as 'The bus is the whole; the window is part of the whole' and 'The school is the whole; the hall is part of the whole'. When this was clear, I was able to formalise this understanding by using representations to show $\frac{1}{2}$. Applying conceptual variation was useful for children to make connections between the concept here. Again, the stem sentence of 'The whole is divided into two **equal parts**; one of the parts is shaded' allowed children to consider what the concept of **half** meant (and also what it did not mean).

It can be useful to **unitise** fractions – using the language of $\frac{3}{4}$= **three one-quarters**. The vocabulary used here reveals the structure of fractions but it does take time to embed, and many examples can be necessary so that children can practise in some depth. It can also help to **count fractions** on number lines as numbers: 'one-ninth, two-ninths, three-ninths…' and then 'one one-ninth, two one-ninths, three one-ninths…' and so on. I find that this helps children later down the learning sequence when unitising for adding or subtracting fractions, because we can say 'three one-ninths add two one-ninths = five one-ninths'. This structure avoids the misconception of adding the **denominators** together, which we sometimes see when tricks and gimmicks are taught.

Representing fractions on a number line and counting forwards (and backwards) can support children's understanding by showing fractions as numbers rather than just as part of a whole. It can also support with **mixed numbers** and **improper fractions** if they are displayed on screen to show children the **equivalences**. These representations may also help children when adding and subtracting fractions (such as $1\frac{2}{4}-\frac{3}{4}=\frac{3}{4}$). We can also use these representations to **compare fractions** using <, > or =. It can also be useful to draw out the generalisation of **unit fractions**: the larger the denominator, the smaller the fraction. This may need other representations to strengthen this, such as bar models and 2D shapes to show that $\frac{1}{2} > \frac{1}{4}$.

We can now consider that if children can find one-quarter of an amount by dividing it into four equal parts, children can make the connection to finding two one-quarters of an amount and three one-quarters of an amount (non-unit fractions). Unitising in this way allows the concept to be accessible for all children, rather than passively accepting 'we divide by the bottom and times by the top', which can become yet another random rule that we use in maths without understanding why.

For **simplifying**, we can teach children that a fraction can be simplified when the **numerator** and denominator have a **common factor** other than 1. We can now make links to **highest common factors** and to other areas of the curriculum, in order for children to be successful with simplifying and **converting fractions**.

This is only a small sample of the vocabulary attached to fractions. Vocabulary matters. How we teach things matters. These things are not separate but part of how we teach effectively. The words that we use impact how we think.

Implementing

From experience, introducing specific and technical maths vocabulary tends to be the area of maths that takes a little more effort to embed and sustain in schools. Many teachers may be unfamiliar with this vocabulary, and you cannot teach what you do not know. Implementing this, therefore, can bring challenges. Having a clear rationale as to why you want to teach this vocabulary can be a good starting point. It may link with enabling children to think mathematically and reason accurately. Once we have a clear rationale around this and we know our 'why', we may then consider how we begin to implement this.

I am going to return now to S-planning. This can be an effective way in which to frame the deliberate implementation of maths vocabulary across school. A way of doing this could be to lay out all of the small steps or learning objectives across a topic in one colour and then systematically map which vocabulary would be needed for each objective and how this could be deliberately planned. Beginning with the vocabulary for place value, the four operations or fractions could be a good place to start, before moving on to other areas of the maths curriculum, such as shape or position and direction.

I often think of vocabulary as the hidden principle of teaching for mastery. To truly master something is to understand and explain it articulately. Teaching key vocabulary at the start of a lesson but then never referring to it or using it throughout your teaching will have a very limited and superficial impact. This is why it is important to ensure that teachers have a solid understanding of the vocabulary to use when teaching maths.

Key takeaways

Vocabulary can be overlooked but it is a fundamental component that supports students' mathematical understanding and reasoning. Precise and systematic teaching of mathematical terminology can significantly enhance students' ability to articulate their thinking and comprehend complex concepts. Vocabulary in mathematics is as essential as it in other subjects like phonics. It needs to be deliberately planned and systematically integrated into lessons. Proper mathematical vocabulary helps to prevent misconceptions and supports the clear communication of mathematical ideas.

Teachers' subject knowledge must be developed in mathematical vocabulary so that they can provide precise explanations and foster deep understanding. Using precise terms like 'addend', 'sum', 'minuend', 'subtrahend', 'product', 'factor', 'dividend', 'divisor' and 'quotient' helps students to grasp the structure of mathematical operations. Procedural variation can help students to see patterns and relationships in mathematics. For example, varying addends in addition problems helps students to understand the commutative property. Systematic use of vocabulary aids in turning routine calculations into reasoning exercises, allowing students to make generalisations and deepen their mathematical thinking.

Teaching terms like 'addend' and 'sum' from a young age helps students to understand the components of addition and make connections across different contexts and number types. Introducing 'minuend', 'subtrahend' and 'difference' clarifies the concept of subtraction and aids in understanding relationships within subtraction problems. Familiarising students with 'factor' and 'product' enables them to generalise that multiplying factors results in a product, and helps in understanding properties like doubling. Terms like 'dividend', 'divisor' and 'quotient' assist in conceptualising division, making it easier for students to grasp inverse relationships between multiplication and division.

The next chapter will delve into how these principles of teaching vocabulary integrate with the broader framework of teaching for mastery, particularly through the lens of the five big ideas: coherence, representation and structure, mathematical thinking, fluency, and variation. Monitoring the mathematics curriculum involves ensuring that these big ideas are embedded in teaching practices, including the deliberate and systematic use of vocabulary. This ensures that students not only learn mathematical concepts but also develop the ability to articulate and reason about these concepts effectively, paving the way for a more profound and holistic mathematical understanding. This transition underscores the interconnectedness of teaching strategies, emphasising that vocabulary is not an isolated component but an integral part of a coherent and structured approach to mathematics education.

As we move forward, we will explore how monitoring the curriculum ensures that these elements work in synergy to achieve teaching for mastery in mathematics.

In the journey towards teaching for mastery in mathematics, the five big ideas serve as the foundational pillars guiding our approach. Each of these ideas plays a crucial role in shaping a comprehensive and effective mathematics curriculum, ensuring that all students develop a deep, connected understanding of mathematical concepts.

8 Monitoring and assessment

Monitoring

Now that your vision is being implemented, it is time to begin to monitor. There are varying ways in which we can monitor, and in my experience in primary, it is something that is not done very well because there can be limited training on how to monitor and what to monitor. This can mean that monitoring is underdeveloped or left to one person (usually the headteacher). It's certainly not something on which I have ever been given any real guidance in my career. With one person responsible for monitoring, it can become a top-down approach, where monitoring is more like checking what people have been doing rather than the impact of what they are doing on learning. Monitoring should be focused on a specific area and should be continuous in its nature. For example, rather than looking for teaching for mastery around the five big ideas, consider focusing on one specific area, such as representations, coherent lesson design, fluency or variation theory. The model in Figure 8.1 may be useful when it comes to ensuring that monitoring is robust.

FIGURE 8.1 A suggested monitoring cycle

Investigate

At this stage, we might have a gut feeling or a hunch, or maybe data or learning walks are suggesting that an area is ripe for development. At this stage, we begin to investigate what the issue may be and potential barriers. We might start to consider particular solutions, programmes or how to develop teaching at this stage. Let's imagine that there is a problem with children at school not making connections between the maths that they are taught. You have visited lessons, spoken to children and looked at data. Books suggest that children complete routine tasks and calculations when number units are taught. Staff mention that time is an issue and they do not always have any capacity to change the activities suggested from the scheme that you use in school. You may reasonably decide on a focus on variation theory, so that children can make connections. You may decide, depending on the size of your school, to begin with particular year groups or people. Here, context and knowing your colleagues are key to success.

Plan

Once we have considered the approach and spoken to key stakeholders, we can then adopt the plan. This should be clear and logical and evaluate the readiness of school, in order to develop the plan well. We might start with professional development through staff meetings, coaching, mentoring and/or teacher research groups, and have a rationale as the impetus for change. Your professional development would cover conceptual variation and procedural variation and the power of teaching using these. It would invite teachers to look at the *structure* of a question and apply variation theory to it in order to develop a carefully sequenced set of questions that deepen thinking. You may provide examples and work with individual teachers on lesson design and the careful design of questions.

Do

Once the plan is in place, we need to monitor how well change is happening. Staff may need support and challenge. Comparing the tasks to which children attend in lessons by looking at books, talking to children and talking to staff will be necessary here. There may be early adopters of the initiative but there will, undoubtably, be those who wait and see whether this is something that is going to be sustained or just another 'fad' from an enthusiastic leader. Revisiting the plan within a term is usually a good call to ensure that the key messages are clear now that everyone has had the opportunity to try the initiative. Delivering more professional development may be appropriate at this stage, or working in small groups or with individuals on designing questions using procedural variation could be time well spent. We want teachers to innovate and use intelligent adoption so that they embed this as part of their teaching each day.

Embed

Monitoring, checking and evaluating the effectiveness and how embedded this initiative is will be key to its success. You may consider how it can be scaled across school. You may ask staff to bring examples of completed tasks to staff meetings and discuss the impact of the task design, including variation theory, or you might do this through a teacher research group or any other effective vehicles that get the conversations moving on teaching and your initiative.

Sustain

When your initiative has legs, you can rest assured that it is part of how maths is taught. I think of this stage as the tweaking stage. If you have prioritised the correct school improvement area, then this stage can last for as long as needed. Take the example of variation: sustaining this way of teaching, across all areas of the curriculum, will take time and effort, with constant revising until it is well established. Once you reach a standard where it is having an impact, you can start the cycle again and investigate the same area or another area that needs developing, and so the cycle continues.

Accountability

A top-level view from leaders is important. They should have an overview of maths in your school and know where the strengths and areas ripe for development lie. Stakeholders who hold you to account will ask about the quality of maths education. Evidence may be collected to triangulate whether what the maths leader is saying is happening in lessons actually is happening in lessons.

A discussion with the maths leader is usual for a maths deep dive. Having your rationale and starting with 'why' can be a good place to begin. Why mastery? Here, you may talk about ambition for *all* learners to understand mathematical structures. This will be followed by lesson visits, discussions with pupils, work scrutinies (often with pupils) and a discussion with the observed teachers. It would be useful for you to have your curriculum, complete with what you want children to know by the end of each year group, printed off. It may also be useful to have a particular stand highlighted to track the coherent nature of your curriculum (place value or multiplication and division tend to work very well for this). Please also make sure that your curriculum includes Early Years!

It can also be useful to have key number fluency facts tracked out (see Chapter 5) so that you can see these tracked across school. Again, this should begin in the Early Years. Some schools call these 'non-negotiables'. Personally, I am not a fan of the phrase. Nothing in education is certain or still. Things change. Thinking changes. Priorities change. Therefore, I think that everything ought to be up for negotiation. I prefer to call this document 'curriculum promises'. These are the facts that children will be systematically taught in school, and the teacher promises to focus on them throughout the year so that children know them and can use them in their maths in their next academic year. Of course, there may be children who still need practice and recapping of the facts. A fluency fact progression document allows teachers to track back to previous years and revisit key fluency facts where necessary. Again, we find ourselves teaching to stage and not age.

Assessment

We need to ask ourselves how we know whether maths is taught well and whether it is assessed well. Are teachers equipped with pedagogical knowledge for the different concepts that they are expected to teach? How is cognitive load managed in class? How adaptive is the teaching to the needs of the children and the responses that they give? How strong is subject knowledge? How is assessment used to check that the curriculum is understood? How are key objectives revisited, reviewed and/or retaught? How do teachers ensure that learning is remembered over time by all? How are speed and accuracy of number facts checked? What is the weighting between formative and summative assessments? How is this information used?

I mentioned the aeroplane analogy earlier in this book, checking the comfort of our passengers at every available opportunity rather than at the end of the journey (or end of a lesson). I'm afraid that I have another aviation metaphor: when we give feedback, we might 'helicopter' around the room to give immediate feedback to children. Notice, here, that I am deliberately not using 'marking', as I think that this often conjures up images of teachers sitting with piles of books after a lesson (or flight) has finished, and so feedback becomes distant and less impactful. In some cases, live marking can inform an intervention that happens on that same day for children who need some additional practice. This can then be coupled with diagnostic assessments and end-of-unit and end-of-term assessments to support thinking around what needs to be retrieved and what interventions may be useful for groups of children.

There can be a well-intentioned focus on end-of-key-stage data to see how many children may be at the expected standard. In some ways, I think that this misses the point of assessment. Time may be better spent identifying the mechanisms for understanding the cause of pupils' lack of progress, as summative assessments do not, by their design, fully assess a whole curriculum. This approach may also lead to teachers and leaders putting their resources and effort into those children on the 'cusp' of the expected standard, at the expense of those children who require the most support.

Early Years

We are born into the world with a natural inclination to learn. Very young children can subitise a small number of objects. By about the age of five or six, children start to realise that more knowledgeable others can support their development. This is why the way in which we introduce children to mathematical ideas in Early Years is so important. We lay the foundations of the big ideas of place value through counting, subitising, the four operations and shape, measure and pattern. A key focus should be on mathematical vocabulary. How is it explicitly planned for across the Early Years curriculum? By the age of five or six, children need help with learning what they do not already know, particularly when introducing maths concepts. Explicit, direct teaching is necessary to achieve this. The environment and provision should allow for opportunities for children to practise what they have been taught. It would be wise to consider any entry gaps in the knowledge of number in particular, and support children in developing these, especially for children from vulnerable backgrounds and disadvantaged groups.

Pupil voice

Pupil voice can be a valuable way of checking how much children know and remember from what they have been taught. Embedding this as part of your monitoring approach can be another way of gathering insights into your curriculum. Here, I'll describe the process that I go through when conducting pupil voice. I often ask for around four to six children and they bring their maths book(s) along with them. We want questions to: be clear, concise and age-appropriate; cover the curriculum, teaching and culture (or all of these); and have a mixture of open and closed questions to illuminate the reality of what happens day to day. Table 8.1 is an example of questions that could be adapted to ask children in maths, alongside what they reveal (curriculum, teaching, culture). Some questions may cross over.

From a practical point of view, I would speak with children in Early Years and Key Stage 1 in their classrooms during a lesson. In Key Stage 2, this may be done in a small group panel set up for this purpose.

TABLE 8.1 Suggested questions and what they could potentially reveal

What does it mean to be a mathematician? *Culture (mindsets)*	Can you show me something in your book that you found tricky? *Culture (mindsets)*	How do you know that you are improving in maths? *Teaching (and assessment)*

Table 8.1 (Continued)

Show me some of the methods of calculation that you have used. *Curriculum*	Find something that you learned recently. Which maths vocabulary did you need for this? *Curriculum*	I can see that you have been learning about ____. What things that you have learned before helped you when you were doing this new work? *Curriculum*
Which resources are the most helpful in lessons? *Teaching*	How does your teacher know that you understand what you have been taught? *Teaching*	How does your teacher help you to remember important maths facts? *Teaching/curriculum*

I will make notes of what children have said and then I triangulate this information with some diagnostic questions that children complete in pairs. These are a series of age-appropriate questions to check whether children can remember what they have been taught in their year group. For place value, with children in Year 6, it may be reading and writing numbers to 10,000,000, identifying the value of digits in numbers up to 10,000,000 and then comparing numbers to 10,000,000 using the inequality symbols. All of this information then allows me to get a bigger picture of the vision and reality in school. Table 8.2 is a simple form that you could populate during pupil voice.

TABLE 8.2 A possible method of recording pupil voice

	Question 1	Question 2	Question 3	Question 4	Question 5	*Questions to see whether children can answer age-related questions (linked to National Curriculum statements) and qualitative information on what children say and do*
Year 1						
Year 2						
Year 3						
Year 4						
Year 5						
Year 6						

After this is complete, you may conduct some thematic analysis to draw out key themes from the pupil voice and consider how these fit in with what you see during learning walks, book looks and discussions with teachers. All of this forms part of your monitoring cycle and can impact on your school improvement plans.

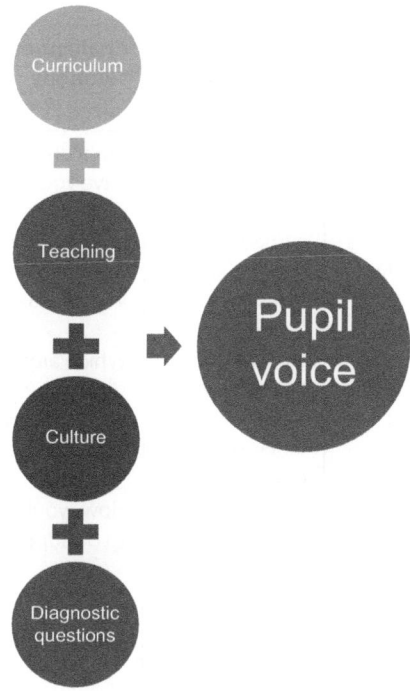

FIGURE 8.2 What can pupil voice tell us?

Staff voice: Gaining insights and enhancing practice

Staff voice is a crucial component of effective monitoring and implementing the five big ideas of teaching for mastery. Engaging staff in reflective dialogue and gathering their insights helps to create a comprehensive picture of teaching practices, challenges and areas for improvement.

Staff voice involves creating structured opportunities for teachers to share their experiences, insights and suggestions regarding their teaching practices and the implementation of mastery strategies. This can take various forms, including individual interviews, focus group discussions, surveys and informal conversations during staff meetings or professional development sessions. The goal is to create an open, supportive environment, where staff feel comfortable sharing honest feedback.

To gain meaningful insights from staff, it is essential to ask well-thought-out questions that cover various aspects of teaching and learning. Table 8.3 gives some examples of questions that can be used around particular themes.

TABLE 8.3 Possible questions to ask staff around themes on which you may want information

Curriculum and pedagogy	Professional development	Assessment	Challenges and solutions	Impact and effectiveness
How do you integrate the five big ideas into your teaching? Can you provide examples of how you use [one of the five big ideas] when designing lessons? Describe a recent successful lesson using the five big ideas as a framework.	What professional development has been the most effective for you in understanding and implementing teaching for mastery? What additional support do you need?	How do progress through the curriculum? How do you assess whether children are making connections between different mathematical concepts? What strategies do you use to ensure that all children are engaged and progressing?	What challenges have you encountered when implementing teaching for mastery in your classroom? Which area of teaching for mastery do you need further development with? How would you like to be supported?	What impact has teaching for mastery and the five big ideas had on your practice? What impact has your teaching had on outcomes for children?

To effectively implement staff voice, begin with careful planning to determine the purpose and scope of gathering staff feedback and decide on the methods to be used, such as surveys or focus groups. Develop clear, focused questions that will yield meaningful insights, perhaps around a whole-school priority or area to develop (or an area that you think is going well in school). Your intention should be to check how accurate your evaluation of maths in your school is. Engage staff by scheduling regular opportunities for them to share their feedback in a supportive environment that encourages open and honest dialogue. Collect and record the feedback systematically, using both qualitative and quantitative methods to capture a comprehensive view. Analyse the data to identify common themes, strengths and areas for improvement, comparing staff feedback with other data sources, such as student voice and lesson observations. Finally, use the insights alongside pupil voice and a look in books to build a clearer picture of maths across the school. This may then inform school improvement plans, develop targeted professional development initiatives based on identified needs, and implement changes in practice, with their impact being monitored over time.

The data gathered from staff voice should be used to drive continuous improvement in teaching practice and student outcomes. Share the findings with staff to validate their input and demonstrate that their voices are heard and valued. Discuss the implications of the data and collaboratively develop action plans. Tailor professional development programmes to address the specific needs and challenges identified through staff voice, fostering a culture of continuous learning and improvement among staff. Regularly revisit the feedback and action plans to assess progress, adjusting strategies as necessary to ensure the effective implementation of mastery teaching practices. By actively incorporating staff voice into the monitoring process, schools can create a more inclusive, reflective and effective approach to teaching for mastery. This not only enhances teaching practices but also fosters a collaborative school culture where continuous improvement is a shared responsibility.

Systems

At the heart of this book, without explicitly mentioning it, is a strong theme of systems. The systems that we used for implementing any initiative are important: they lead to small, incremental changes, which can amount to cultural shifts over time. We may consider the systems in place to develop teachers or teaching assistants in our school, to support their subject knowledge and teaching approaches. From this, systems may further target specific groups of children, such as lower-attaining pupils, the bottom 20 per cent of cohorts or classes, pupils from disadvantaged backgrounds, pupils with SEND or a mixture of all of these, to support all the learners that we can.

Consideration may be given to the deliberate and systematic way in which change happens. We cannot do everything with everyone all at once. We may want to consider rationales for where we might start professional development for teachers, based on the children in each class. For example, starting in Early Years or Year 1 can allow for strong beginnings to then embed across Key Stage 2, in the terms or years ahead. In doing so, our monitoring becomes more focused on pupils' thinking and the quality of the practice that they undertake.

Key takeaways

It is essential to reflect on the journey and continuously consider the next steps to ensure the successful implementation of the five big ideas of teaching for mastery. Throughout this chapter, we have explored the importance of monitoring as a continuous and integral part of the teaching process. Effective monitoring goes beyond mere compliance; it is about understanding the impact of teaching practices on student learning and making informed decisions to support all stakeholders in your school(s).

The 'investigate, plan, do, embed' cycle provides a robust framework for monitoring. Starting with investigation, we identify areas ripe for development, supported by data, observations and conversations. Planning involves clear, logical steps with professional development at their core, ensuring readiness and coherence. The implementation phase ('do') focuses on supporting and challenging staff, fostering innovation and intelligent adoption of new practices. Embedding these practices requires ongoing monitoring and evaluation to ensure that the initiative is firmly rooted in daily teaching and then sustained.

Pupil voice has been highlighted as a crucial component of monitoring, offering valuable insights into what children know and remember. By engaging with pupils through structured questions and diagnostic assessments, we can triangulate this information with observations and teacher discussions, providing a comprehensive view of the curriculum's impact.

The role of strong systems in implementing initiatives and driving improvements cannot be overstated. Systematic and deliberate changes lead to significant cultural shifts over time. Effective systems support teachers and teaching assistants, targeting specific groups of children and ensuring that all children are supported. Monitoring these systems focuses on pupil thinking and the quality of practice, ensuring that initiatives are deeply embedded and impactful.

Conclusion: Teaching until mastery

So there we have it. Fourteen years of experience (at the time of writing) in teaching in schools and reflecting on mathematics, its culture, strategies, pedagogy, language and all that it encompasses, have been distilled into the pages of this book. Teaching until children have mastered concepts is the main message here. Ensuring high success and application of skills is essential for children to succeed.

A common thread throughout this book has been subject knowledge. Teacher subject knowledge is essential to effective teaching. Combining this with pedagogical prowess is the sweet spot for which we all strive. Essentially, knowing what you are teaching and how to teach it is our key aim. This needs substantial time and effort over a sustained period to ensure that it is effective. Investing in continuous professional development is crucial for this endeavour.

It is rather unfortunate that the chapters in this book are discrete. The intention of teaching until mastery is for all of these ideas to be happening in every lesson simultaneously. A well-thought-out, coherent lesson that uses representations to reveal underlying mathematical structures is a good starting point. In turn, this allows children to think mathematically and have the vocabulary to reason accurately, developing fluency as they go. Careful questioning and depth of understanding can be achieved through intelligent practice and variation theory, so that all children have an accessible starting point and can make connections.

Having worked in a range of different schools, those schools that have a clear vision and execute it well tend to be the most successful. They have professional fidelity to teaching for mastery and the strategies described in this book. They are relentless in their ambition for teachers, teaching assistants and children. They want everyone to understand mathematics by moving away from superficial tricks and gimmicks to unlock a world of understanding. They revisit their curriculum plans regularly, focus on what is important and always look for ways in which to improve. Walking around these schools and seeing the approach in action – the deliberate and informed choices that teachers are making – always makes me smile. You can get a strong sense of how well-embedded teaching for mastery is after a learning walk, and we know that effective teaching does not happen by mistake.

I have seen first-hand the transformational impact of teaching for mastery in schools. When we assume that children can, we open up ambition and higher expectations for them to achieve well. We dedicate ourselves to excellence and hold our values and beliefs to account. We do away with the myth that learning happens in neat, lesson-shaped segments and welcome the idea that children forget and need help remembering. Children may need to see concepts many times, and we may need to reteach certain elements of our curriculum from previous academic years to ensure that children understand. We meet children at their stage and not age.

At the heart of these pages is a key message about challenging conceptions of ability. If we assume that children cannot, then they probably will not. If we try to teach the curriculum in a way that scaffolds learning into small steps and has high ambition at the centre of our thinking, then we raise the potential for all children. We do not want to inadvertently or unconsciously limit children's understanding of maths. The 'fleas in a jar' experiment taught us that. Culture eats strategy for breakfast.

By reading this book, you have already demonstrated your commitment to improvement. You may already have a clear rationale for change, for what is going well and what you may need to develop in your school(s). Of course, we cannot do everything all at once. I encourage you to think carefully and deeply about an area on which to focus. What do all of your different methods of monitoring tell you? Be clear on why you want to improve some areas and celebrate other areas done with greater effect. Is it lesson design and coherence, or do you want staff to give closer attention to the representations that they are using to teach mathematical concepts? How will you do this and what will you do to measure the impact?

It can be useful to visualise what you want your curriculum and teaching to look like in three years' time and work backwards from there. What are the small steps that you can take each term (short-term goals)? What will this look like across an academic year (medium-term goals)? And what will success look like after a few years (long-term goals)? Imagine that you started this year; the children in Early Years to Key Stage 1 would have a very different experience by the end of Key Stage 2. Of course, we cannot be fixed in our thinking. We must respond to any emerging needs or priorities as we go, while keeping our gaze firmly on the long-term goal.

Implementing teaching for mastery is not without its challenges. Teachers often face obstacles such as limited resources, varying levels of student ability and resistance to change. However, these challenges can be overcome with strategic planning and a commitment to professional development. For instance, schools can allocate time for collaborative planning and sharing of best practices. Providing ongoing support through coaching and mentoring can help teachers to feel more confident and capable in their roles.

As we reach the end of this book, I urge you to take the insights and strategies discussed and put them into action. Begin with small, manageable changes and gradually build on them. Engage with your colleagues, share your experiences and learn from each other. Remember that the journey towards teaching for mastery is ongoing and requires dedication, reflection and a willingness to adapt. Your commitment to improving mathematics education will make a significant difference in the lives of children, helping them to achieve their full potential.

References

Back Matter (1978), *Philosophical Transactions of the Royal Society of London. Series A, Mathematical and Physical Sciences*, 288, no. 135. Available at: www.jstor.org/stable/74950 [Accessed 13 June 2024].

Christie, C. (2017), 'Developing fluency in addition and subtraction facts', Number Sense Maths. Available at: https://numbersensemaths.com/media/1841/achieving-fluency-in-addition-and-subtraction-facts-article.pdf [Accessed 17 July 2024].

Department for Education (DfE) (2013), 'The National Curriculum in England: Key Stages 1 and 2 framework document'. Available at: https://assets.publishing.service.gov.uk/media/5a81a9abe5274a2e8ab55319/PRIMARY_national_curriculum.pdf [Accessed 11 July 2024].

Department for Education (DfE) (2020), 'Mathematics guidance: Key Stages 1 and 2: Non-statutory guidance for the National Curriculum in England'. Available at: https://assets.publishing.service.gov.uk/media/6140b7008fa8f503ba3dc8d1/Maths_guidance_KS_1_and_2.pdf [Accessed 9 July 2024].

Department for Education (DfE) (2021), 'National Curriculum in England: Mathematics programmes of study'. Available at: www.gov.uk/government/publications/national-curriculum-in-england-mathematics-programmes-of-study/national-curriculum-in-england-mathematics-programmes-of-study [Accessed 15 July 2024].

Department for Education (DfE) (2024), 'Early Years Foundation Stage statutory framework'. Available at: https://assets.publishing.service.gov.uk/media/65aa5e42ed27ca001327b2c7/EYFS_statutory_framework_for_group_and_school_based_providers.pdf [Accessed 17 July 2024].

Ebbinghaus, H. (1885), *Über das Gedächtnis: Untersuchungen zur Experimentellen Psychologie*. Leipzig: Duncker & Humblot.

Gray, E. and Tall, D. (1993), *Success and Failure in Mathematics: Procept and Procedure*. Coventry: Mathematics Education Research Centre, University of Warwick. Available at: www.davidtall.com/eddiegray/92b-procepts-primary.pdf [Accessed 15 August 2025].

National Centre for Excellence in the Teaching of Mathematics (NCETM) (n.d.), 'Exemplification of ready-to-progress criteria'. Available at: www.ncetm.org.uk/classroom-resources/exemplification-of-ready-to-progress-criteria [Accessed 18 July 2024].

National Centre for Excellence in the Teaching of Mathematics (NCETM) (2017), 'Five big ideas in teaching for mastery: The fundamental characteristics that underpin teaching for mastery in all school and college phases'. Available at: www.ncetm.org.uk/teaching-for-mastery/mastery-explained/five-big-ideas-in-teaching-for-mastery [Accessed 5 July 2024].

National Centre for Excellence in the Teaching of Mathematics (NCETM) (2022), 'The essence of mathematics teaching for mastery'. Available at: www.ncetm.org.uk/teaching-for-mastery/mastery-explained/the-essence-of-mathematics-teaching-for-mastery [Accessed 13 June 2024].

National Centre for Excellence in the Teaching of Mathematics (NCETM) (2024a), 'Mastering number at KS2'. Available at: www.ncetm.org.uk/maths-hubs-projects/mastering-number-at-ks2 [Accessed 13 June 2024].

National Centre for Excellence in the Teaching of Mathematics (NCETM) (2024b), 'Primary mastery professional development'. Available at: www.ncetm.org.uk/teaching-for-mastery/mastery-materials/primary-mastery-professional-development [Accessed 13 June 2024].

Number Sense Maths (2024), 'Number Sense Maths: Programmes and approach'. Available at: https://numbersensemaths.com [Accessed 13 June 2024].

Ofsted (2021), 'Research review series: mathematics'. Available at: https://www.gov.uk/government/publications/research-review-series-mathematics/research-review-series-mathematics [Accessed 13 June 2024].

Ofsted (2023), 'Subject report series: Maths'. Available at: www.gov.uk/government/publications/subject-report-series-maths [Accessed 13 June 2024].

Rosenshine, B. (2012), *Principles of Instruction: Research-Based Strategies That All Teachers Should Know. American Educator*, Spring 2012. Washington, DC: American Federation of Teachers. Available at: https://www.aft.org/sites/default/files/Rosenshine.pdf [Accessed 28 July 2025].

Outlaw, F. (1977), *What They're Saying*. Greenville News, 2 May.

Russell, S. J. (2007), 'Developing computational fluency with whole number in elementary grades'. Available at: https://investigations.terc.edu/inv2/wp-content/uploads/2017/10/Developing-Computational-Fluency-with-Whole-Numbers-in-the-Elementary-Grades.pdf [Accessed 13 June 2024].

Sharples, J, Albers, B. and Fraser, S. (2019), 'Putting evidence to work: A school's guide to implementation: Guidance report', Education Endowment Foundation. Available at: https://dera.ioe.ac.uk/id/eprint/31088/1/EEF-Implementation-Guidance-Report.pdf [Accessed 9 July 2024].

Sharples, J., Eaton, J. and Boughelaf, J. (2024), 'A school's guide to implementation', Education Endowment Foundation. Available at: https://educationendowmentfoundation.org.uk/education-evidence/guidance-reports/implementation [Accessed 13 June 2024].

Skemp, R.R. (1986), *The Psychology of Learning Mathematics*, 2nd ed. London: Penguin.